Vibrational Spectra of Electron and Hydrogen Centers in Ionic Crystals

D. Bäuerle

Contents

1. Introduction

The fundamental role played by lattice vibrations in nearly all fields of solid state physics, be it in connection with absorption of elastic or electromagnetic waves, superconductivity, or ferroelectricity, has stimulated a growing interest in lattice vibrational spectra of pure and perturbed crystal lattices. Such vibrations can be studied directly be means of inelastic neutron scattering which is to be prefered for the study of bulk vibrational modes, and by infrared and Raman spectroscopy which, because of their inherent sensitivity to relatively low defect concentrations, are excellent techniques for investigating the defect induced response of a crystal lattice. Such defects, which may be foreign atoms, ions, or molecules, but which may also be dislocations, voids, and so on, are interesting because they lower the point symmetry of the lattice, thereby relaxing the selection rules for photon-phonon interaction processes, and thus allowing the optical activation of (somewhat perturbed) host lattice vibrations which are inactive in the unperturbed crystal.

It is the aim of this work to give a survey on the infrared and Raman spectra of some color centers in ionic crystals, particularly for those whose mass is very small when compared with the mass of the host lattice ions. Electron, hydrogen, and vacancy centers, defects which have this property, are of special interest for the clarification of lattice dynamical problems. The reason is that the atomic and electronic structure of these centers is well established from their electronic absorption spectra, and from the analysis of electron-paramagnetic resonance (EPR), and/or electron nuclear double resonance (ENDOR) data. Furthermore, some lattice dynamical properties of these centers were already known. They were derived indirectly from the shape of the electronic absorption lines which show strong temperature – dependent broadening due to electron – phonon coupling. Extensive investigations of this kind were performed for the electronic absorption of F centers in alkali halides. In addition, electron and hydrogen centers can be transformed by simple and directly controlable photochemical reactions, thus allowing one to study, in a well defined way, point defects of various potentials, charges, and point symmetry. The experimental and theoretical analysis of the infrared and Raman spectra of these defects yields detailed information on the site symmetry and local bonding of the defect, and on the host latticeband modes.

The main features observed in the infrared and Raman spectra of the special defects discussed in this work are representative of the vibrational properties of defective crystal lattices in general.

2. General Remarks

2.1. Formation, Structure, and Electronic Absorption of Electron and Hydrogen Centers

The electron and hydrogen centers in alkali halides whose vibrational properties will be investigated in this work are shown schematically in Fig. 1. The following remarks on the production, atomic structure, and the electronic absorption of these centers are of an introductory nature. For more details, refer to special literature on this topic [1–4].

The most well known electron center is the F center, which is a trapped electron on an anion lattice site. In alkali halides, F centers can be produced by heating the crystal in alkali vapor (additive coloration), by electron injection (e.g. electrolytic coloration), by X-irradiation or by a photochemical reaction from U centers (see Fig. 1). F centers give rise to an electronic absorption in the visible spectral region (in KBr near 2.06 eV at $T = 4.2$ K). Irradiating with light into the visible F band at low temperatures results in the formation of F' centers (two electrons in an anion vacancy) and anion vacancies (α centers). The electronic absorption of F' centers is located in the visible to near infrared region (in KBr near 1.77 eV at 170 K) while α centers absorb in the near ultraviolet (in KBr near 6.2 eV at 90 K). Irradiating the F band with visible light at room temperature leads to an association of F centers into pairs (M centers), trimers (R centers), or even more complex aggregates. The

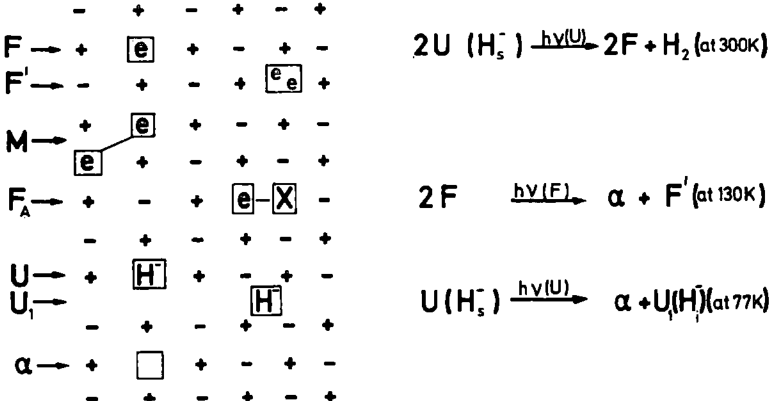

Fig. 1. Left side: Structural models for some electron and hydrogen centers in alkali halides. F center: one electron in an anion vacancy. F' center: two electrons in an anion vacancy. M center: two neighboring electrons in [110] or equivalent directions. F_A center: one electron associated with foreign alkali ion of smaller size. $U(H_s^-, D_s^-)$ center: substitutional hydrogen ion. $U_1(H_i^-, D_i^-)$ center: interstitial hydrogen ion. α center: anion vacancy. Right Side: Photochemical reactions. Energy $h\nu$ refers to electronic bands

electronic absorption of these centers is located in the visible to near infrared region. An F center associated with one foreign alkali ion of smaller size in the first shell is called an F_A center [in analogy to Fig. 32, case (b)].

Since the point symmetry in the F_A configuration is reduced to C_{4v}, the threefold degenerate first excited state of the F center splits twofold. F_A centers therefore give rise to two electronic transitions which are located near the F center transition (in KBr for F_A(Na) 2.07 and 1.90 eV, respectively at 4 K).

Negative hydrogen ions placed substitutionally on anion sites are called U centers, because they give rise to an electronic transition in the near ultraviolet region (in KBr near 5.51 eV at $T=6$ K). U centers can be produced by treating additively coloured crystals under hydrogen atmosphere at about 200 K below the melting point, or by doping the melt with alkali hydride. U centers can be converted by UV or X-irradiation at room temperature into F centers and interstitial hydrogen molecules (see Fig. 1). Irradiating at liquid nitrogen temperature yields interstitial hydrogen centers, U_1 centers, and anion vacancies (generally this process leads simultaneously to the production of a small amount of F centers according to the first reaction shown in Fig. 1). The electronic U_1 band is very broad even at low temperatures (maximum at about 4.6 eV in KBr at 55 K).

In alkaline earth fluorides, defects which possess the structure of the F center can be produced by X-ray irradiation or by additive coloration when heating samples in the presence of aluminium metal (at about 800 K). Treating additive colored samples under hydrogen atmosphere (at about 1100 K) results in the production of substitutional hydrogen ions [5]. The formation of substitutional hydrogen atoms by photochemical reactions has been established as well [6]. The production of interstitional hydrogen ions and hydrogen atoms was established in rare earth ion doped CaF_2 (see Sections 3.1 and 3.3) [7].

2.2. Selection Rules for Infrared and Raman Spectra

In pure, ideal crystals the interaction of elementary excitations with light demands conservation of (quasi-) momentum, This follows from the translational symmetry of crystal lattices. Because the wavelength of light is very large when compared with the crystal unit cell, only lattice vibrations with wave vector $k \approx 0$ can be directly excited. In ionic crystals vibrations near the center of the Brillouin zone are dipole active if they belong to transverse optical branches in the phonon dispersion curves. The corresponding oscillators are sometimes called "dispersion oscillators" or "reststrahl oscillators". Dispersion oscillators give rise to optical absorption in the middle to far infrared frequency region.

When point defects are introduced into the crystal, the lattice vibrations can no longer be classified according to the wavevector, but only according to the point symmetry of the defect. For the interaction with light the k conservation law is no longer valid, i.e. all phonons may become infrared active in the *one* phonon absorption. The corresponding dipole moment is determined by the changed eigenvectors and the additional dynamical charge of the defect.

From atomic physics it is well known that optical transitions will be allowed only for certain combinations of quantum states. The selection rules can be expressed most easily in group theoretical terms (for details see Refs. [8–11]). Assume that the lattice dynamical problem is described by a Hamiltonian H which is invariant under operations $\{R\}$ of a group G, so that

$$RHR^{-1} = H . \tag{1}$$

The vibrational ground state, the excited state, and the dipole operator each transform like some irreducible representation Γ_m, Γ_n, and Γ_D of the group G. Then one has to form the direct product of any of these irreducible representations, e.g. $\Gamma_m \times \Gamma_D$. If

$$\Gamma_m \times \Gamma_D \in \Gamma_n \tag{2}$$

Table 1. Notations for irreducible representations of point group O_h

M	L	BSW	B	vdLB	HJ
A_{1g}	Γ_1^+	Γ_1	Γ_1^+	α	Γ_s
A_{2u}	Γ_2^-	$\Gamma_2{}'$	Γ_2^-	β	Γ_f^1
A_{2g}	Γ_2^+	Γ_2	Γ_2^+	β'	Γ_i
A_{1u}	Γ_1^-	$\Gamma_1{}'$	Γ_1^-	α'	Γ_k
E_g	Γ_{12}^+	Γ_{12}	Γ_3^+	γ	Γ_d^1
E_u	Γ_{12}^-	$\Gamma_{12}{}'$	Γ_3^-	γ'	Γ_h
T_{2g}	Γ_{25}^+	$\Gamma_{25}{}'$	Γ_5^+	ε	Γ_d^2
T_{1u}	Γ_{15}^-	Γ_{15}	Γ_4^-	δ	Γ_p
T_{1g}	Γ_{15}^+	$\Gamma_{15}{}'$	Γ_4^+	δ'	Γ_g
T_{2u}	Γ_{25}^-	Γ_{25}	Γ_5^-	ε'	Γ_f^2

The notation in the first column is according to Mulliken [12]. The following notations are due to Loudon [8], Bouckaert et al. [13], Bethe [14], von der Lage and Bethe [15], and Howarth and Jones [16]. Indices "g", "u" and +, − refer to even and odd parity respectively. Because $O_h = T_d \times I$ this table is also valid for point group T_d when symbols "g", "u", and +, − are omitted.

the matrix element may be nonzero, i.e. the transition is allowed by group theory. If

$$\Gamma_m \times \Gamma_D \notin \Gamma_n$$

the matrix element *must* be zero and the transition is forbidden.

For the case of a cubic lattice (point group O_h) the vibrational ground state transforms like A_{1g} while the dipole operator transforms according to T_{1u} (in the following the notation of Mulliken [12] is used; for convenience other notations are listed in Table 1). Therefore, one finds from Eq. (2) that the matrix element

$$(A_{1g}|T_{1u}|\Gamma) \tag{3}$$

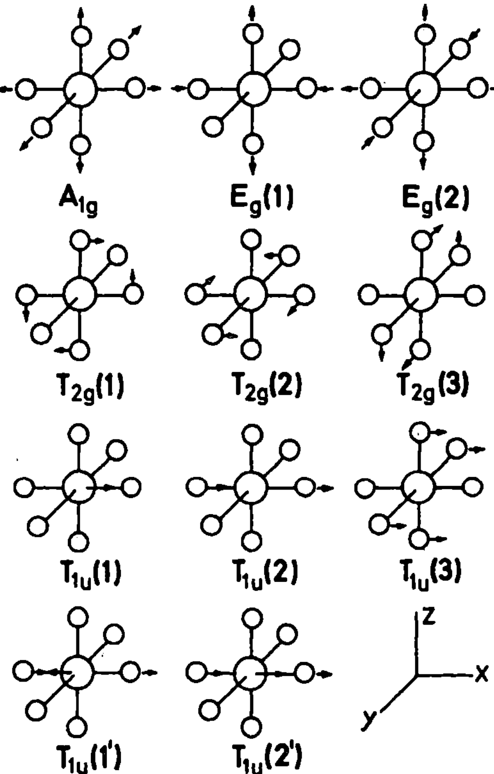

Fig. 2. Illustration of irreducible configurations of an impurity with octahedral symmetry (after Ref. [20]). They correspond to the symmetry vectors of the first nearest neighbor (1 nn) coordinates as introduced in Section 2.4. Configurations A_{1g} and E_g are nondegenerate and twofold degenerate, respectively. Configurations T_{2g} and T_{1u} are threefold degenerate. Linear combinations of $T_{1u}(1)$ and $T_{1u}(2)$ configuration yield configurations $T_{1u}(1')$ and $T_{1u}(2')$. The latter are mainly used in this work. For convenience the prime will, in the future, be omitted

may be non-zero only if $\Gamma \equiv T_{1u}$. The T_{1u} configurations are shown in Fig. 2. Only these will contribute to the one phonon infrared absorption.

While in pure cubic crystals with inversion symmetry (point group O_h) first order Raman scattering is forbidden, in perturbed crystals even-parity oscillations of the nearest neighbors of a defect atom at a site of inversion symmetry may become Raman active. The modes of A_{1g}, E_g, and T_{2g} symmetry contribute to the first order defect induced Raman scattering. This can easily be verified from the conclusions above. The mechanism for scattering has its origin in the modulation of the electronic polarizability by the motion of the impurity and its neighbors. The polarizability can be described by virtual dipolar transitions from low-lying to higher electronic states. Such a process is of third order, and one can write the transition amplitude in the form

$$\sum_{i,j} \frac{(0, A_{1g}|\Gamma_D|i, A_{1g})\,(i, A_{1g}|H_{EP}|j,\Gamma)\,(j,\Gamma|\Gamma_D|0,\Gamma)}{(E - E_{i,A_{1g}}) \cdot (E - E_{j,\Gamma})} \tag{4}$$

where 0, and i,j refer to the electronic ground state and the dipole active excited states, respectively. H_{EP} is the electron phonon interaction and transforms like A_{1g}. A_{1g} refers to the phonon ground state while Γ refers to the phonon states excited in the Raman experiment. Γ_D is the dipole operator which transforms like T_{1u}. Then, it is easily seen that Eq. (4) satisfies the condition for non-vanishing matrix elements [Eq. (2)], only if

$\Gamma_D \times \Gamma_D \in \Gamma$.

Because

$$T_{1u} \times T_{1u} = A_{1g} + E_g + T_{1g} + T_{2g} \tag{5}$$

the only phonons which can couple to the (virtual) electronic transition are of A_{1g}, E_g, T_{1g}, and T_{2g} symmetry. If the Raman tensor is symmetric

Table 2. First-order selection rules for Raman scattering from point defects in cubic lattices

| | Incident field polarization | |
	[100]	[110]
Scattered field polarization		
$\parallel$	A_{1g}, E_g	A_{1g}, T_{2g}
$\perp$	T_{2g}	E_g

(which is the case for non-resonant Raman scattering) only phonons of symmetry A_{1g}, E_g, and T_{2g} are involved.

The contribution of each symmetry type to the Raman spectrum can be determined from three independent measurements. The types of modes contributing to the spectra for various orientations of incident and scattered electric field polarizations are given in Table 2.

By comparing Eqs. (4) and (5) one finds that in cubic crystals with inversion symmetry (point group O_h) infrared active vibrations are Raman inactive and vice versa. Therefore, infrared absorption and Raman scattering complement each other.

2.3. Localized Modes and Resonant Modes

Ionic crystals have one or more dispersion oscillators (see Section 2.2) whose frequencies are given approximately by the frequencies of the optically active eigenvibrations of the crystal unit cell. Alkali halides which have cubic symmetry (point group O_h) and which have two atoms per unit cell have only one dispersion oscillator, which gives rise to strong optical absorption in the middle to far infrared frequency region. At liquid helium temperature, the crystals are tansparent outside the reststrahl absorption.

The introduction of point defects into a crystal may lead to an activation of optically inactive band vibrations, i.e. to defect induced absorption within the phonon band regions (see Section 2.2 and 4.1), and to eigenvibrations outside the unperturbed band modes. The defect induced band mode absorption is generally very broad and essentially reflects the density of (more or less perturbed) host lattice phonon states weighted by the strength of coupling to the light. In special cases, however, relatively sharp, distinct absorption lines may occur also. This can happen if the defect and/or its nearest neighbors are vibrating at a frequency where the host lattice phonon density is small and/or if this vibration is strongly decoupled from the sourrounding lattice due to strongly reduced local force constants. Such vibrations are called resonant modes or quasilocalized modes. In ionic crystals resonant modes are in general observable only within the acoustic band region, because the optical band is covered by the strong reststrahl absorption.

Eigenvibrations outside the unperturbed band modes of the host lattice are called localized modes. The band width of localized modes is in general very small because direct harmonic damping is impossible. The lifetime of the localized oscillator is determined by indirect coupling to lattice modes. This coupling may be due to anharmonic terms or due to higher order dipole moments in the potential (see Section 3.1.2).

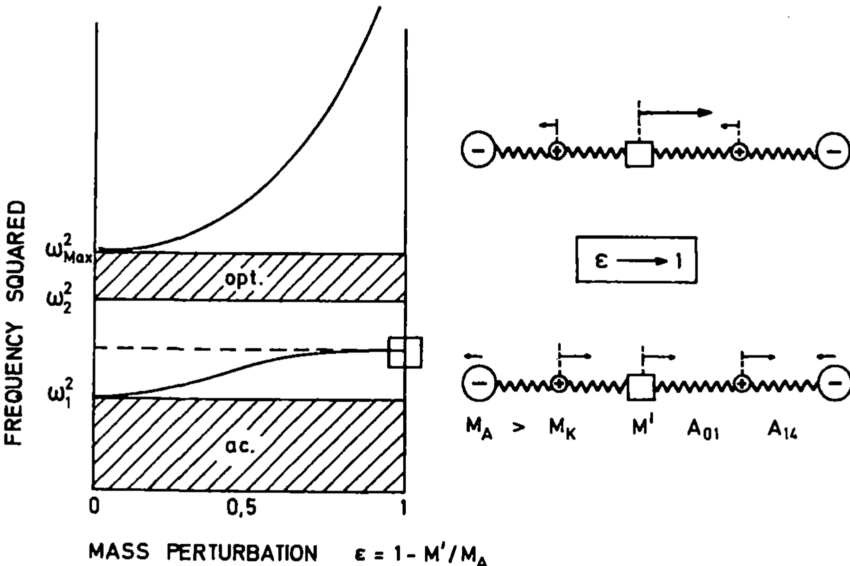

Fig. 3. Localized mode and in-gap mode frequencies as function of substitutional defect mass, M', in the linear diatomic chain approximation with $A_{01} = A_{14} \cong A$. A refers to the unperturbed force constant of the host lattice. M_K and M_A refer to host lattice cations and anions, respectively. Displacements are shown for the limiting case $M' \ll M_A$, i.e. $\varepsilon = 1 - M'/M_A \to 1$

In alkali halides localized modes may be located either above the optical band modes or within the acoustic-optical band gap. In the latter case these modes are often called "in-gap modes". The situation can well be demonstrated for the example of defects whose mass, say M', is small when compared with the mass of the host ions, as it is the case e.g. for substitutional hydrogen centers. One gets insight into the problem most easily if one solves a linear diatomic chain model corresponding e.g. to the motion of atoms in the x-direction of Fig. 2 when transverse force constants are neglected. Assuming the defect at the center, one obtains for the case of pure mass perturbation (force constants unchanged) the result of Fig. 3. The main features are two infrared active modes: firstly a high frequency localized mode which splits from the top of the optical band and whose frequency Ω_L approaches infinity for vanishing defect mass, and, secondly, an in-gap mode which splits from the top of the acoustic band and whose frequency stays finite even for $M' = 0$.

The high frequency localized mode is of the configuration T_{1u} (1) in which the defect atom moves out of phase with the first nearest neighbors (see Figs. 2 and 3). This vibration is similar to the reststrahl oscillation where the lattice of positive ions vibrates against the lattice of negative ions. The only difference is that the amplitude is very large at the defect site and decreases exponentially with the distance from the center. It is obvious from this model that the frequency of the $T_{1u}(1)$ vibration strongly depends also on the force constant A_{01} between the defect and first nearest neighbors (1 nn). This force constant may be changed with respect to the host lattice force constant due to a change in the overlap between the defect atom and 1 nn. In fact, this situation can be studied at a $T_{1u}(1)$ vibration which was found for substitutional hydrogen centers in alkali halides. It will be further discussed in Chap. 3.

The in-gap mode has the configuration $T_{1u}(2)$. The displacements of the defect atom and its neighbors are shown for the limiting case $\varepsilon \to 1$ in Fig. 3 (see also Fig. 2). In this vibration the 1 nn move in phase with the defect. The amplitude is not localized at the defect site but is about the same at 1 nn ions. Such a vibration was also found for substitutional hydrogen centers. However, it turned out that the corresponding frequency is not located in the phonon gap but in the upper half of the acoustic band. This frequency shift back into the acoustic band is an effect of the local force constant changes which have to be taken into account. It directly follows from the simple model of Fig. 3 that the mode frequency will sensitively react to force constant changes between 1 nn and fourth nearest neighbors (4 nn). In this model changes in force constants between the defect and 1 nn have no influence at all in the limit $\varepsilon \to 1$, because then the defect and 1 nn vibrate as a "rigid" oscillator against 4 nn. In Chap. 4, $T_{1u}(2)$ type in-gap modes and resonant modes associated with color centers as mentioned in Section 2.1 will be discussed. Examples of infrared active in-gap modes and resonant modes due to defects in polar and homopolar crystals in general are given in reviews by Genzel [17], and by Sievers [18].

The introduction of defects into a crystal can lead also to defect induced Raman scattering from (somewhat perturbed) host lattice band modes, and to Raman active localized modes and resonant modes. When the defect atom is at a site of inversion symmetry only the even-parity oscillations of the nearest neighbors are Raman active (see Section 2.2). This case was studied in connection with F centers in alkali halides. It will be discussed in Chap. 5. When the defect atom is *not* at a site of inversion symmetry, the motion of the impurity itself may be Raman active in a localized or a resonant mode. The incident (monochromatic) light of frequency ω_i is then scattered to frequencies $\omega_i \pm \omega_L$ or $\omega_i \pm \omega_R$, respectively. This case was not yet observed for centers discussed in this work, however, it could eventually occur for interstitial hydrogen centers which are discussed in Section 3.2.

2.4. Theoretical Background

In this section only a few remarks on the theory of infrared absorption and Raman scattering of crystal lattices with point defects are made.

The theory of infrared absorption of crystal lattices with point defects was reviewed by several authors as e.g. by Benedek and Nardelli [19], and by Klein [20]. The linear absorption coefficient is, neglecting reflection, defined in terms of the intensity of the transmitted light relative to the incident light, I/I_0, and is given by

$$\alpha = (1/d)\ln(I/I_0) = (4\pi\omega/cn(\omega))\,\mathrm{Im}\,\chi(\omega) \tag{6}$$

where d is the sample thickness, ω the frequency of incident light, c is the velocity of light and $n(\omega)$ is the refractive index. The dielectric susceptibility χ is a tensor in general. For cubic crystals one has $\chi_{\alpha\beta} = \chi \cdot \delta_{\alpha\beta}$. The susceptibility is originally obtained from a linear response theory. Using the Green function formalism one can write

$$\chi = (NZ^2/v_0\mu) \lim_{\varepsilon\to 0} (\varphi_{TO}|[G_0^{-1} + p\,G_0^{-1}\,TG_0]^{-1}|\varphi_{TO})$$

$$= (NZ^2/v_0\mu) \lim_{\varepsilon\to 0} [(\omega_{TO}^2 - (\omega + i\varepsilon)^2)^2 + p(\varphi_{TO}|T|\varphi_{TO})]^{-1} \tag{7}$$

where N is the number of unit cells, and v_0 the cell volume. Z is the effective charge, μ the reduced mass, and p the probability of finding one defect in the unit cell. The T matrix is given by

$$T = V(I + G_0 V)^{-1}$$

where G_0 is the Green function for the unperturbed lattice. V is the perturbation matrix which contains the perturbation in mass and force constants. V and T are generally defined in a $3rN$ dimensional vector space (r is the number of particles in the unit cell). However, if the perturbed region within the lattice is sufficiently localized, the elements of V and T are non-zero only in a $3q$-dimensional subspace, the so called "impurity space" (q is the number of ions affected by the perturbation). ω_{TO} is the frequency of the dispersion oscillator with $k\approx 0$. φ_{TO} is the corresponding eigenvector.

To simplify the calculation of the T-matrix elements one can introduce special symmetry vectors (shell vectors), which are described in more detail in Refs. [21] and [22]. These symmetry vectors constitute a complete set of basis functions and are most suitable for describing the lattice vibrations around a defect with a finite range of perturbation.

In a lattice with point-symmetry group G the lattice ions may be assumed to be arranged on "shells" around the defect. A "shell" is defined by all lattice points which transform into one another under all symmetry operations of G. Symmetry vectors are those basis vectors to irreducible representations of G, the amplitudes of which are nonzero only on a single shell. The symmetry vectors are described by four indices. s is the shell number, starting with $s = 0$ for the defect itself, Γ gives the irreducible representation of G, r is the index of multiplicity of Γ in the shell s, and j is the index of degeneracy. For the projections $s(\Gamma, j, s, r)$ of the eigenvectors φ_{T0} one obtains

$$s(\Gamma, j, s, r) \equiv (\varphi_{T0} | \sigma(\Gamma, j, s, r))$$

and in analogy

$$\hat{T} = (\sigma | T | \sigma') = (\sigma | V (I + G_0 V)^{-1} | \sigma')$$

$$= \sum_{\sigma''} (\sigma | V | \sigma'') (\sigma'' | R | \sigma') \equiv \hat{V}\hat{R} \, . \tag{8}$$

For calculating χ one can substitute the second term in Eq. (7) by $(s | \hat{T} | s)$.

Assuming that the translational symmetry of the effective charges is *not* disturbed in the perturbed lattice, one can write the matrix of effective charges in the form

$$Z = I \times Z_r, \qquad \text{where } Z_r = e \begin{pmatrix} +1 \\ -1 \end{pmatrix} .$$

As a consequence only those infrared-active modes which can be deduced from T_{1u} modes of the unperturbed lattice will contribute to the absorption coefficient in the perturbed lattice, i.e., $s \neq 0$ only if $\sigma = \sigma(T_{1u}, j, s, r)$. Hence

$$(s| = (s(T_{1u}), 0| \, .$$

This results in a further simplification of the projected perturbation matrix $\hat{V}$ and the Green function $\hat{G}$. The scalar product $(s | \hat{T} | s)$ has the simple form

$$(s | \hat{T} | s) = (s(T_{1u}) | \hat{V}(T_{1u}) [I + \hat{G}(T_{1u}) \hat{V}(T_{1u})]^{-1} | s(T_{1u})) \, . \tag{9}$$

To find the only unknown in Eq. (9) one has to discuss the vector $(s(T_{1u})|$ and the projected perturbation matrix $\hat{V}(T_{1u})$ for special models. This will be done only in Section 4.2.3 for the case of infrared absorption of F centers. For other cases the reader is refered to the original literature.

The specific force constant model to be used in the following sections is shown in Fig. 4. Changes in force constants between the defect and 1 nn, $A01 = (A_{01} - A)/A$, and between 1 nn and 4 nn, $A14 = (A_{14} - A)/A$, are taken into account. If $\Delta k = 0$ this model corresponds to the extended model by Gethins et al. [23].

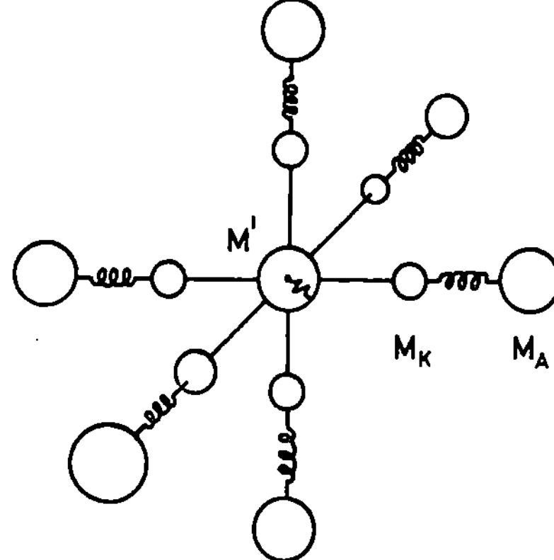

Fig. 4. Force constant model essentially discussed in this work. $-\!-\!- \; A_{01}$ is the force constant between the defect and 1 nn. $\sim\!\sim\!\sim A_{14}$ between 1 nn and 4 nn. $\sim\!\sim\!\sim k$ is the shell-core spring

The theory of Raman scattering from crystals with point defects was developed by Sennett [24], Xinh [25] Maradudin [26], Benedek and Nardelli [27], and some others.

The relation between the intensities of the incident light of frequency ω_i and the scattered light of frequency ω_s can be described by the fourth-order tensor

$$i_{\alpha\gamma\beta\lambda}(\Omega) = (2\pi)^{-1} \int dt \, e^{-i\Omega t} (P_{\beta\lambda}(t) P^*_{\alpha\gamma}(0)) \tag{10}$$

where $\Omega = \omega_i - \omega_s$ is the energy transfer, and $P_{\beta\lambda}(t)$ is the time dependent operator for the electronic polarizability tensor of the crystal. One now can expand the polarizability in series of lattice ion displacements, say $u(l, k)$ (l being a Bravais vector, and k a cell index). The zero order term (no phonons involved) accounts for Rayleigh scattering ($\Omega = 0$), while the first and higher order terms account for first and higher order Raman scattering, respectively. The first order process is characterized by the coefficient.

$$P_{\alpha\beta,\mu} \equiv \partial P_{\alpha\beta} / \partial u_\mu(l, k) \,.$$

As it was shown by Maradudin [26] the first order part of Eq. (10) can be written for the Stokes process in the form

$$i_{\alpha\gamma\beta\lambda}(\Omega) = (p\hbar/\pi M)(n(\Omega) + 1) \sum_{\substack{l,k,\mu \\ l',k',\mu'}} P_{\beta\lambda,\mu}(l,k) \, P_{\alpha\gamma,\mu'}(l',k')$$

$$\times \operatorname{Im}(l,k,\mu|[I + G_0 V]^{-1} G_0|l',k',\mu') \tag{11}$$

where the same notations as above have been used. M is the mass of 1 nn of the defect. In analogy to the treatment given above one now can introduce symmetry vectors in order to simplify the calculation of Eq. (11).

For cubic symmetry there are only three independent components of the Raman tensor

$$i_{zzzz} = i_{11}; i_{xxyy} = i_{12}; i_{xyxy} = i_{44}.$$

For a perturbation extending only to 1 nn of the defect, there are three different non-zero terms

$$P_{xx,xx} \equiv \tfrac{1}{3}(\alpha + 2\beta); P_{xx,yy} \equiv \tfrac{1}{3}(\alpha - \beta); P_{xy,xy} \equiv \gamma.$$

Then the components of the first order Raman tensor for cubic crystals are

$$i_{11} = (p\hbar/6\pi M \Omega)(n(\Omega) + 1)[\alpha^2 \hat{\varrho}(A_{1g}) + 2\beta^2 \hat{\varrho}(E_g)]$$

$$i_{12} = (p\hbar/6\pi M \Omega)(n(\Omega) + 1)[\alpha^2 \hat{\varrho}(A_{1g}) - \beta^2 \hat{\varrho}(E_g)] \tag{12}$$

$$i_{44} = (p\hbar/2\pi M \Omega)(n(\Omega) + 1)|\gamma^2 \hat{\varrho}(T_{2g})|$$

where $\hat{\varrho}(\Gamma)$ is the projected one-phonon density

$$\hat{\varrho}(\Gamma) = 4M\Omega \operatorname{Im}(\sigma|[I + G_0 V]^{-1} G_0|\sigma').$$

The coefficients α, β, γ can be obtained experimentally from uniaxial stress experiments. For the case of F centers they were determined by Gebhardt and Maier [28] for various alkali halides from the stress splitting of the electronic F band.

3. Infrared Vibrational Absorption: High Frequency Localized Modes

High frequency localized modes occur well above the maximum host lattice frequency. Such modes were found for substitutional hydrogen ions, $U(H_s^-, D_s^-)$ centers, in a number of compounds. Extensive experimental and theoretical investigations were performed in alkali halides ([30–49], and [22, 23, 29, 40, 50–64] respectively), and in alkaline earth fluorides ([5, 65–67], and [5] respectively). The results will be reviewed in Section 3.1. High frequency localized modes due to interstitial hydrogen ions, $U_1(H_i^-, D_i^-)$ centers, were studied experimentally in a number of alkali halides [69–72], and in CaF_2 [7]. The only theoretical investigation was performed for the case of $KBr:H_i^-$ [61]. Section 3.2 reviews the results. High frequency localized mode absorption due to interstitial hydrogen atoms in LiF [73, 74], and in CaF_2 [75] is fairly well established. The results are mentioned in Section 3.3.

With respect to electron centers like F centers, Rosenstock and Klick [29] have argued that the frequency of the electronic transition $1s-2p$ could be explained by a classical "mechanical" vibration of the electron. The model used by these authors is linear, and does not include local force constant changes. For KCl they obtained for the "F center localized mode frequency" about 3.2 eV, while the measured value of the $1s-2p$ transition is about 2.3 eV at 4.2 K. A much better approximation can be obtained when calculating the F center localized mode frequency from the experimentally determined main line frequency for the H_s^- center localized mode by taking into account only the mass perturbation, i.e. by assuming that the local force constants, in particular A_{01}, are the same for the electron and the hydrogen ion (see also Section 4.2.4). One then obtains values which reproduce the frequency of the visible F band within 5–20% (for KCl one estimates about 2.4 eV). However, such a mechanical model of the electronic transition is not in particular useful.

The investigations reported in this chapter should serve to give an idea about the information which can be obtained from optical investigations of defective crystal lattices. In particular, it is shown how local site symmetries of hydrogen centers in various compounds can be derived from the number of vibrational transitions observed. The frequency positions and intensities of lines allow conclusions on the local oscillator potential, and on local vibrational amplitudes respectively. The investigation of temperature dependent line widths and frequency shifts, and of additional absorption structures observed, yields information on the nature and the amount of phonon-phonon coupling, and on host lattice phonon densities. The results may apply to pure host lattices as well, i.e. the structure and damping of host lattice dispersion oscillators may be discussed along similar lines.

3.1. U Centers (H_s^-, D_s^- Centers)

3.1.1. The Main Line Frequency

Fig. 5 shows a typical localized mode absorption spectrum for the example of H_s^- centers in KBr. This absorption is located in the near infrared spectral region. At room temperature the absorption shape is broad and structureless. Sidebands, which are located nearly symmetrically to the main line at 445 cm^{-1} are observed with decreasing temperature. The main line is due to the motion of the hydrogen ion in a cubic

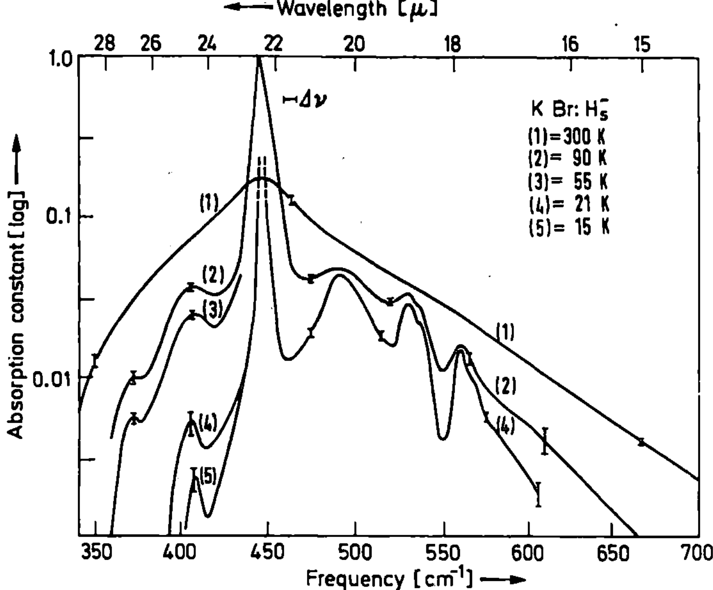

Fig. 5. Infrared absorption of the H_s^- center localized mode in KBr (after Ref. [37])

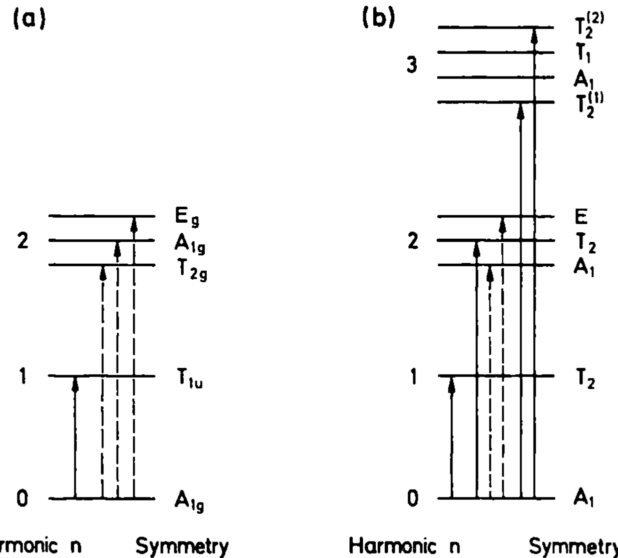

Fig. 6. (a) Energy level scheme of an anharmonic oscillator in a cubic (O_h) environment corresponding to U centers in alkali halides. Full arrow corresponds to an infrared active transition. Transitions indicated by dotted arrows are only Raman active. (b) Same as (a) but for T_d symmetry. Transitions observed in the infrared are shown by full arrows. Dotted arrows indicate infrared transitions which are induced by shear strain (after Ref. [82])

potential. The line is threefold degenerate and corresponds to a vibrational transition from an A_{1g} ground state $(n=0)$ to a $T_{1u}(1)$ excited state $(n=1)$. This transition is shown in Fig. 6 (a). The transition energy is listed in Table 3 for various alkali halides. A model for the $T_{1u}(1)$ vibration is shown in Fig. 2 of Section 2.2

For a phenomenological description of the main line frequency one can treat the hydrogen ion as moving in a *static* potential of O_h symmetry. This approach is physically reasonable since, during one period of oscillation of the "heavy" neighbor ions in a (perturbed) host crystal band mode, the "light" hydrogen ion makes many complete oscillations, so that during any of them, it sees the surrounding ions as essentially at rest. This strong localization of the U center vibration may be directly seen from the isotope shift in the frequency which is about $\sqrt{2}$ for H_s^- and D_s^- centers (see Table 3). Such a value is expected from the mass ratio for "harmonic" H_s^- and D_s^- oscillators moving in a rigid box. In this approximation 1 nn of the hydrogen are held fixed (compare Fig. 4).

A further support of this simple model follows from the localized mode absorption intensities which are related to the vibrational displacements. One expects the integrated absorption for the H_s^- localized

Table 3. Main line frequency and halfwidth for U center localized mode in alkali halides, and alkaline earth fluorides[a]

Substance	Frequency $\nu_L(H_s^-)$ [cm^{-1}]	Halfwidth $\Delta\nu_L(H_s^-)$ [cm^{-1}]	Frequency ratio $\nu_L(H_s^-)/\nu_L(D_s^-)$	Temperature [K]	References
^{6}LiF	1030.9			20	[18]
LiF	1027	4.2	1.38	20	[38]
	1015	18	1.37	300	[44]
NaF	859.5		1.40	20	[38]
	846.7	12	1.39	300	[44]
NaCl	563	5.6	1.38	90	[30, 37]
	565	43		300	[37]
NaBr	498	17	1.38	90	[30, 37]
	504	60		300	[30, 37]
NaI	426.8		1.34	10	[41]
	430.6			100	[31]
KF	725.5			100	[31]
KCl	502	2.3	1.40	90	[30, 32, 33, 37]
	497	32	1.38	300	[37]
KBr	446	6	1.40	90	[30, 32, 33]
	444			300	[37, 40]
	886.3 $(S. T_{2g})$		1.40	11	[76] R
	894.8 $(S. A_{1g})$			11	[76] R
	904.5 $(S. E_g)$		1.40	11	[76] R

Table 3 (continued)

Substance	Frequency $v_L(H_s^-)$ [cm^{-1}]	Halfwidth $\Delta v_L(H_s^-)$ [cm^{-1}]	Frequency ratio $v_L(H_s^-)/v_L(D_s^-)$	Temperature [K]	References
KI	382	14.5	1.37	90	[30, 31, 37]
	392			300	
	753.1 (S, T_{2g})		1.39	8	[76] R
	759.9 (S, T_{2g})			8	[76] R
	767.9 (S, E_g)		1.40	8	[76] R
RbF	703.1			100	[31]
RbCl	476	4.8	1.40	90	[30, 37]
	466		1.38	300	[30, 37]
RbBr	425	8		90	[30, 37]
	412			300	[30, 37]
RbI	360			70	[30]
CsCl	425		1.40	20	[45]
	417			300	[45]
CsBr	364.1	0.4	1.42	6	[36, 49]
	363	35	1.40	300	[36, 49]
CsI	282.8	0.6	1.41	7	[45, 49]
	303		1.38	300	[45, 49]
CaF$_2$	965.6	<0.7	1.39	20	[5]
	969	5.0		16	[77] R
	957.8	8.7	1.39	290	[5]
	1895 (S, A_1)	<4.0		16	[77] R
	1894.0 ,,			77	[78] SI
	1919.8 (S, T_2)		1.39	20	[5]
	1923 ,,	<4.0		16	[77] R
	1945.0 (S, E)			77	[78] SI
	2912.2 (T, T_2)		1.39	20	[5]
	2825.6 (T, T_2)		1.38	20	[5]
SrF$_2$	893.2	1.7	1.39	20	[5]
	896	4.8	1.40	16	[77] R
	1747 (S, A_1)	<4	1.39	16	[77] R
	1775.9 (S, T_2)		1.39	20	[5]
	1778 ,,	<4	1.39	16	[77] R
BaF$_2$	806.6	8.8		20	[5]
	805	2.2	1.40	16	[77] R
	798.2	14		290	[5]
	1566 (S, A_1)	<4.2	1.38	16	[77] R
	1596.2 (T, T_2)			20	[5]
	1597 ,,	<4.2	1.39	16	[77] R

* If not indicated further, values refer to infrared active fundamentals. Higher harmonic transitions were included in the Table for convenience. Their values follow parantheses. First symbol in parantheses, i.e. S or T, refers to second harmonic ($n = 2$), and third harmonic ($n = 2$) frequencies respectively. Second symbol in parantheses refers to the symmetry of the excited vibrational state. R after reference means measured by Raman scattering; SI after reference means stress induced transition.

mode to be twice that of the D_s^- mode. Experimentally a ratio of 1.92 is found [33, 43].

Calculating the main line frequency on the basis of the simple model discussed, and assuming *pure* mass perturbation (setting the force constant A_{01} equal to the host lattice force constant A) one obtains values which are about 40% to 60% too high when compared with experimental values as listed in Table 3. This indicates that a considerable lowering in local force constants has to be taken into account. In fact, this was confirmed in more sophisticated calculations using Green function techniques. On the other hand, these calculations did show also that for a description of further details of the localized mode absorption, force constant changes at least up to 4 nn have to be taken into account. It is returned to this point in Section 3.1.4. In spite of the usefulness of the simple model described above, a rigorous treatment of the U center vibrational problem requires the dropping of both assumptions: the complete localization, and the harmonic motion of the hydrogen oscillator. The motion of the neighboring ions will influence the main line frequency for H_s^- and D_s^- centers differently. This is due to the different degree of localization for H_s^- and D_s^- vibrations (a measure of the degree of localization is, roughly speaking, the distance of the main line frequency from the maximum host lattice frequency). In fact, the deviation from the value $\sqrt{2}$, observed for the frequency ratio for H_s^- and D_s^- vibrations, may be explained in part by including the motion of neighboring ions. On the other hand, the "large" deviation observed in this frequency ratio, cannot be explained in the harmonic approximation alone. Note that for NaI : H_s^-, D_s^- one finds, e.g., $v_L(H_s^-)/v_L(D_s^-) = 1.34$! This indicates that fairly large anharmonic contributions to the harmonic localized mode frequency may be present as well.

In the following the simple model discussed above will be extended by including anharmonic terms in the oscillator potential, which is still assumed to be static. Expanding the potential energy in a power series in the displacements of the hydrogen ion from equilibrium one can write the Hamiltonian in the form

$$H = \tfrac{1}{2} M' \Omega_L^2 r^2 + C_1(x^4 + y^4 + z^4) + C_2(x^2 y^2 + x^2 z^2 + y^2 z^2) + \ldots \qquad (13)$$

where M' is the hydrogen mass and $r^2 = x^2 + y^2 + z^2$. Because the anharmonic terms are expected to be small when compared with the harmonic term the unperturbed wave functions are those of a spherically symmetric harmonic oscillator, having energy levels

$$E_n = \hbar \Omega_L(n + \tfrac{3}{2}). \qquad (14)$$

These equally spaced degenerate levels are split and shifted by the anharmonicity (see Fig. 6). In O_h symmetry, where cubic terms in Eq. (13) are absent, the first excited state $n = 1$ is only shifted. Because of the different vibrational amplitude for H_s^- and D_s^- centers, this anharmonic shift is of different size for H_s^- and D_s^- levels. It explains in part the deviation from the harmonic frequency ratio. The sixfold degenerate $n = 2$ level is split by the anharmonicity into levels of A_{1g}, E_g, and T_{2g} symmetry. The corresponding transitions from the A_{1g} ground state ($n = 0$) to the $n = 2$ excited states are "only" Raman active and will be discussed in Section 5.1. The quartic terms in Eq. (13) allow two infrared active transitions from the ground state to excited states with $n = 3$. For H_s^-, D_s^- centers in alkali halides these transitions were not yet found. Recently, Akhvlediani and Politov [73] correlated a line observed at $1900\ \mathrm{cm}^{-1}$ in neutron irradiated $LiF : OH^-$ crystals with a third harmonic transition of T_s^- (tritium) centers. This correlation, however, seems somewhat arbitrary; furthermore, the high oscillator strength of this transition is not understood from own experiments with H_s^- and D_s^- centers. In particular, Dötsch [80] could not observe a third harmonic transition in LiF and NaF even with H^- center concentrations up to 1 Mol%. However, before discussing the influence of anharmonicity further, we shall turn briefly to the results in alkaline earth fluorides and rare-earth trifluorides.

In alkaline earth fluorides the U center has tetrahedral site symmetry (T_d). One therefore has to add to the expansion of Eq. (13) a term of the form $Bxyz$. As a result, in the energy level scheme shown in Fig. 6 (b) a transition between the ground state ($n = 0$) and the second excited state ($n = 2$) (full arrow) becomes infrared active[1]. Second and third harmonic transitions were observed in $CaF_2 : H_s^-, D_s^-$ in $SrF_2 : H_s^-, D_s^-$ and in $BaF_2 : H_s^-$ by Elliott et al. [5]. The corresponding frequencies are listed in Table 3. Transition frequencies obtained from Raman scattering [77], and from experiments with uniaxial stress [78] are included in the Table for convenience. The linewidths should be viewed with some caution, since they are markedly concentration dependent for the high defect center concentrations used. It should be mentioned that experiments with tunable CO_2 lasers did enable one to study the step-wise excitation for each of the three $n = 2$ levels [81].

The measured transition frequencies can be used to calculate the anharmonic coefficients Ω, B, C_1, and C_2. In this connection Elliott et al. [5] did estimate the effect of anharmonic terms in the Hamiltonian by perturbation theory. The energy for the perturbed vibrational levels relative to the unperturbed ground state can be written in the form

$$E_n(\Gamma) = nh\Omega_L + (\hbar/(2M'\Omega_L))^2\,(C_1\mu_1(\Gamma) + C_2\mu_2(\Gamma))$$
$$- B^2\lambda(\Gamma)\,\hbar^2/(24\,M'^3\Omega_L^4) \tag{15}$$

---

[1] One could, on the other hand, argue that the observation of a second harmonic indicates that in alkaline earth fluorides the local site symmetry of the hydrogen ion is T_d.

where μ_i and λ are tabulated for representations Γ under consideration in Table 4. For the case of O_h symmetry one has to set $B \equiv 0$. Values for Ω, B, C_1, and C_2 as obtained from Eq. (15) and the measured transition frequencies of Table 3 are listed in Table 5. The corresponding values for U centers in KBr and KI are included. They are based on Raman scattering experiments (see Chap. 5). With the anharmonic coefficients of Table 5 one can calculate from Eq. (15) the complete energy level schemes of Fig. 6.

The existence of localized modes due to hydrogen and deuterium ions in the rare-earth trifluorides LaF_3, CeF_3, PrF_3, and NdF_3, was reported by Jones and Satten [79]. The infrared spectra show two strong absorption lines [for LaF_3 e.g. at $818\ cm^{-1}$ and at $1168\ cm^{-1}$ (4.2 K)], which are probably due to hydrogen ions on fluorine sites with D_3 symmetry. The substitution of deuterium results in frequencies which are reduced by a factor of 1.4. These modes were also established from the

Table 4. Values $\mu_i(\Gamma)$ and $\lambda(\Gamma)$ for anharmonic localized oscillator in cubic environment

Harmonic n	Symmetry Γ	$\mu_1(\Gamma)$	$\mu_2(\Gamma)$	$\lambda(\Gamma)$
0	A_1	9	3	1
1	T_2	21	7	5
2	A_1	45	15	21
	E	45	9	3
	T_2	33	15	13
3	A_1	45	27	25
	T_1	57	21	15
	$T_2^{(1)}$	$\begin{pmatrix} 57 & 0 \\ 0 & 81 \end{pmatrix}$	$\begin{pmatrix} 25 & 2\sqrt{6} \\ 2\sqrt{6} & 15 \end{pmatrix}$	$\begin{pmatrix} 27 & 6\sqrt{6} \\ 6\sqrt{6} & 13 \end{pmatrix}$
	$T_2^{(2)}$			

Table 5. Coefficients for anharmonic oscillator potential for U centers in alkali halides and alkaline earth fluorides

Substance	$\Omega_L(H_s^-)$ [cm^{-1}]	$\Omega_L(D_s^-)$ [cm^{-1}]	B $\times 10^{12}$ [erg/cm^3]	C_1 $\times 10^{19}$ [erg/cm^4]	C_2 $\times 10^{19}$ [erg/cm^4]	Temperature [K]	References
KBr	444.6	318.2	0	0.986	−2.30	10	[76] R
KI	377.8	272.3	0	0.576	−1.356	8	[76] R
CaF$_2$	981.1	702.1	7.87	−1.01	−2.32	20	[82]
SrF$_2$	907.4	*	6.20	−2.19	−1.80	20	[78]
BaF$_2$	827.2		3.98	−8.5	−1.7	20	[78]
	817		3.36	−1.35	3.06	4.2	[77]

* Approximately one can set $\Omega_L(D_s^-) \approx \Omega_L(H_s^-)/\sqrt{2}$.

vibronic spectra. In addition to the two intense lines, four absorption lines with smaller intensity were observed. These additional lines may be due to hydrogen ions on other sites which are occupied to a lesser extent. A strong correlation of these lines is not yet possible. One difficulty is that the crystal structure of the rare-earth trifluorides is not completely clarified. If the structure belongs to the space group D_{3d}^4, there are three fluorine sites which have D_3, C_3, and C_1 symmetry, respectively.

3.1.2. Multiphonon Processes

The coupling of the U center localized mode to lattice modes becomes evident in the temperature dependent halfwidth and frequency shift of the main line, and in the occurence of phonon sidebands. It was shown by Bilz et al. [51] and by Elliott et al. [5] that for the case of U centers in alkali halides and alkaline earth fluorides this coupling is mainly of anharmonic nature. Contributions due to higher-order dipole moments are small. For a complete description of the observed anharmonic effects further terms which contain combinations of H_s^- ion displacements and displacements of ions on nearby lattice sites must be added to Eq. (13). Expanding the potential energy again up to terms of fourth order and transforming to phonon creation and annihilation operators, one can write the Hamiltonian for the localized oscillator in the form

$$H = H_0 + H_A \tag{16}$$

where H_0 is the harmonic part of the Hamiltonian. The anharmonic part can be written in the form

$$\begin{aligned}
H_A = &\frac{1}{3!} \sum_{L,i,j} V_3(L,i,j)(a_L + a_L^+)(a_i + a_i^+)(a_j + a_j^+) \\
&+ \frac{1}{4!} \sum_{L,i,j,k} V_4(L,i,j,k)(a_L + a_L^+)(a_i + a_i^+)(a_j + a_j^+)(a_k + a_k^+) + \dots
\end{aligned} \tag{17}$$

Indices i, j, k refer to lattice phonons. The influence of H_A on the localized mode can be made plausible by introducing diagrams as shown in Fig. 7. These symbolize transitions between quantum states. Straight and bent lines are introduced for illustrating phonon-phonon (anharmonic) interactions. Single lines refer to lattice phonons. Double lines refer to the localized oscillator. At vertices (intersection points) one has to connect a number of lines corresponding to the order of anharmonicity (three for third order anharmonicity via V_3, and four for fourth order anharmonicity via V_4). Closed diagrams with *one* vertex yield a contribution to the self energy in *first* order, diagrams with n vertices in n th

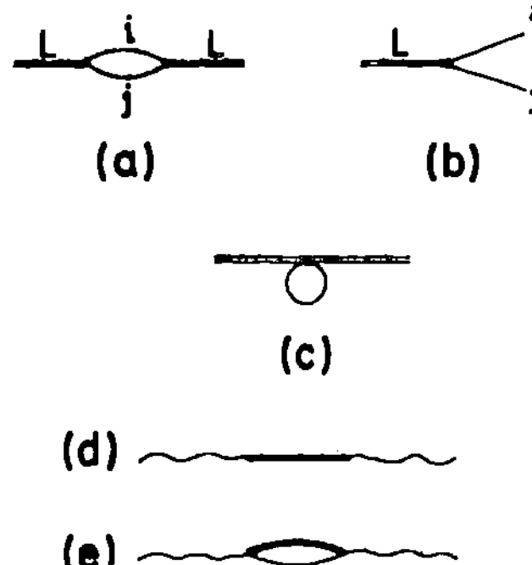

Fig. 7. Self energy diagrams illustrating phonon-phonon (anharmonic) interactions [cases (a – c)], and photon-phonon interactions [cases (d), (e)]. Double lines refer to the localized oscillator (Ω_L), single lines to lattice phonons ($\omega_i, \omega_{j...}$). (a) Localized oscillator frequency shift $\Delta_L(\omega)$ due to two phonon decay into virtual intermediate states through V_3^2 (L, i, j). (b) Imaginary part of self energy, characterizing damping due to two phonon decay. This diagram is obtained by bisecting diagram (a). Interaction through V_3^2 (L, i, j). (c) Frequency shift through coupling to phonon ω_j. Interaction through $V_4(L, L, i, j)$. – Diagrams (a)-(c) may follow processes (d) and (e). (d) Resonance (undamped) between photon and localized oscillator Ω_L. Interaction through first order dipole moment M_1. (e) Same as case (d) but for $\omega = \Omega_L \pm \omega_i$. Interaction through second order dipole moment M_2. This process is negligible for U centers in alkali halides and alkaline earth fluorides

order. For example the process in Fig. 7 (a) is associated with V_3^2, and (c) with V_4. Oscillating lines illustrate photons. They describe in connection with straight or bent lines photon-phonon interaction.

Fig. 7 (a) shows a typical third order contribution to H_A: An intermediate state with two excited phonons is coupled by V_3^2 (L, i, j) to the localized mode, resulting in a shift of the harmonic oscillator frequency Ω_L. Diagram (b) is obtained by bisecting diagram (a). It represents the imaginary part of the self energy, and describes the damping due to decay into two arbitrary band phonons ω_i and ω_j. All contributions of this type are obtained by summing over all ω_i and ω_j within the phonon bands. Of course, energy must be conserved in these processes, i.e. $\Omega_L = \omega_i \pm \omega_j$. A different type of a closed diagram which is associated with fourth order anharmonicity is shown in Fig. 7 (c). In this case the localized oscillator interacts virtually with one phonon. This process is independent of frequency. It therefore contributes to the shift but not to the damping.

Diagrams as introduced in Fig. 7 can be used for a systematic description of all contributions to the complex susceptibility. They can be built up from the simple diagrams (a)–(c) in Fig. 7. A survey on the most important contributions in connection with the U center localized mode is given in Table 6. A more complicated process which is discussed by

Table 6. Contributions to anharmonic frequency shift and damping. (s) refers to sum processes and (d) refers to difference processes

Frequency shift $\Delta_L(\omega)$	Damping $\Gamma_L(\omega)$		$\Delta_L(\omega)$	$\Gamma_L(\omega)$	Temperature dependence $\Delta_L(\omega)$, $\Gamma_L(\omega)$
1.		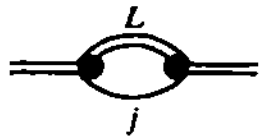  high frequency sideband (s)	$-\dfrac{2(\Omega_L \pm \omega_i)}{(\Omega_L \pm \omega_i)^2 - \omega^2}$	$\delta(\Omega_L \pm \omega_i - \omega)$	$n_L + n_i + 1 \approx n_i + 1$ (s)
		 low frequency $\sim$ (d)			$-n_L + n_i \approx n_i$ (d)
2.		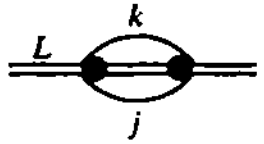  2-phonon side-bands (s)	$-\dfrac{2(\Omega_L + \omega_i \pm \omega_j)}{(\Omega_L + \omega_i \pm \omega_j)^2 - \omega^2}$	$\delta(\Omega_L + \omega_i + \omega_j - \omega)$	$(n_L + 1)(n_i + 1)(n_j + 1)$ $-n_L n_i n_j \approx (n_i + 1)(n_j + 1)$ (s)
		 halfwidth (H_s^- and D_s^-)		$\delta(\omega_i - \omega_j)$	$3[(n_L + 1)(n_i + 1)n_j$ $-n_L n_i(n_j + 1)] \approx 3(n_i + 1)n_j$ (d)
3.		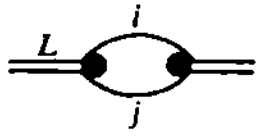  halfwidth (D_s^-)	$-\dfrac{2(\omega_i \pm \omega_j)}{(\omega_i \pm \omega_j)^2 - \Omega_L^2}$	$\delta(\omega_i \pm \omega_j - \Omega_L)$	$n_i + n_j + 1$ (s)
4.		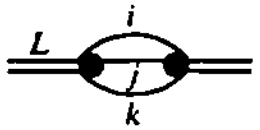  halfwidth (H_s^- and D_s^-)	$-\dfrac{2(\omega_i + \omega_j \pm \omega_k)}{(\omega_i + \omega_j \pm \omega_k)^2 - \Omega_L^2}$	$\delta(\omega_i + \omega_j \pm \omega_k - \Omega_L)$	$(n_i + 1)(n_j + 1)(n_k + 1)$ $-n_i n_j n_k$ (s)
5.		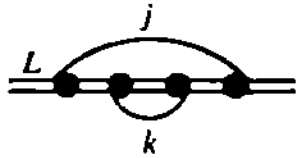  halfwidth (H_s^- and D_s^-)	$-\dfrac{2(\Omega_L + \omega_j - \omega_k)}{(\Omega_L + \omega_j - \omega_k)^2 - \Omega_L^2}$	$\delta(\omega_j - \omega_k)$	$3[(n_L + 1)(n_j + 1)n_k$ $-n_L n_j(n_k + 1)] \approx 3(n_j + 1)n_k$ (d)
6.		$-$	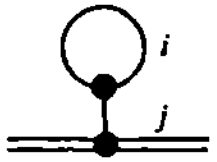 $-1/\omega_j$	0	$2n_i + 1$
7.		$-$	 $+1$	0	$2n_i + 1$

Elliott et al. [5] is shown in the fifth row of Table 6. A complete survey on the anharmonic processes is given by Bilz et al. [51].

Because the term H_A in Eq. (16) is small when compared with H_0, the contributions of multiphonon processes to the energy and damping of the localized oscillator can be calculated by perturbation theory. For the diagram in the third row of Table 6 the first term of Eq. (17) for H_A yields, in lowest order, the linewidth[2]

$$2\Gamma(\omega_L) = 36\pi \sum_{i,j} |V_3(L,i,j)|^2 [(\bar{n}_i + 1)(\bar{n}_j + 1) - \bar{n}_i\bar{n}_j] \delta(\Omega_L - \omega_i - \omega_j) \quad (18)$$

and the frequency shift

$$\Delta(\omega_L) = 36 \sum_{i,j} \frac{|V_3(L,i,j)|^2 (\omega_i + \omega_j)}{\Omega_L^2 - (\omega_i + \omega_j)^2} [(\bar{n}_i + 1)(\bar{n}_j + 1) - \bar{n}_i\bar{n}_j]. \quad (19)$$

Here difference processes for phonons ω_i and ω_j are not included. Further results for the shift function $\Delta_L(\omega)$ and the damping function $\Gamma_L(\omega)$ are listed in the third and fourth column of Table 6. Only constant factors and the anharmonic coefficients $|V_n(L, i \ldots)|^l$ have been omitted. It is understood that $\Delta_L(\omega)$ and $\Gamma_L(\omega)$ are not independent of one another. They are real and imaginary parts of a damping function

$$\Pi_L(\omega) = \Delta_L(\omega) + i\Gamma_L(\omega)$$

and are related by the Kramers-Kronig-relations.

The temperature dependence of $\Delta_L(\omega)$ and $\Gamma_L(\omega)$ is listed in the fifth and sixth column of Table 6. It can be directly calculated by using the properties of boson operators a^+ and a.

$$a_i^+ \; |n_i) = (n_i + 1)^{1/2} |n_i + 1)$$
$$a_i \; |n_i) = n_i^{1/2} |n_i - 1). \quad (20)$$

Here one has to take into account all processes in thermal equilibrium corresponding to the same matrix element. $\bar{n}_i = (\exp(\hbar\omega_i/kT) - 1)^{-1}$ is the average thermal occupation number of phonons ω_i. In order to get net transition probabilities one has to subtract in Eq. (18) and (19) multiphononcreation and -annihilation processes. In the following, multiphonon processes and their connection with the temperature dependent halfwidth and the sideband structure of the localized mode main line will be discussed.

---

[2] $\Gamma(\omega)$ can not always be directly compared with Δv, the measured line width, since other effects such as strain broadening may come into play.

3.1.3. Temperature Dependence of the Main Line

The temperature dependence of the *halfwidth* of the main line is shown for the typical example of KCl with H_s^- and D_s^- centers in Fig. 8. For temperatures < 50 K the lifetime of the localized mode is determined by the decay into band modes. Because

$$\omega_L(D_s^-) < 2\omega_{max} < \omega_L(H_s^-) \tag{21}$$

(where ω_{max} is the highest band mode frequency; see Appendix), the D_s^- oscillator can decay into two band modes by $V_3(L,i,j)\ a_L a_i^+ a_j^+$ (diagram 3 in Table 6). For the temperature dependence of the two phonon decay process one obtains from Eq. (20)

$$(\bar{n}_i + 1)(\bar{n}_j + 1) - \bar{n}_i \bar{n}_j \approx 2\bar{n} + 1$$

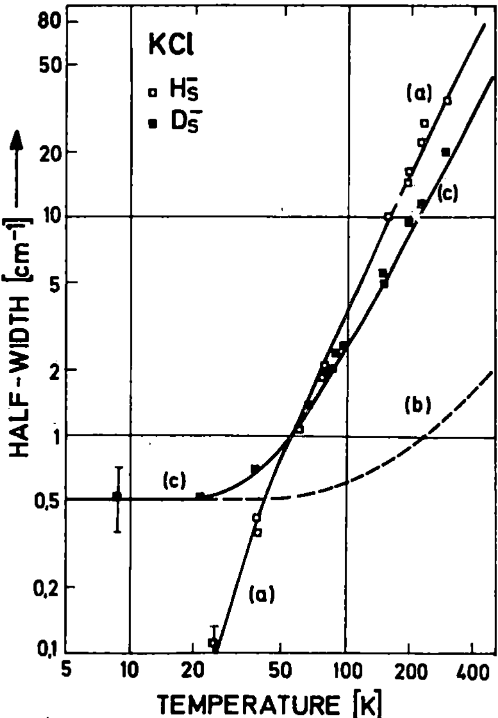

Fig. 8. Temperature dependence for the halfwidth of the main line. Curve *a*: Theoretical fit for H_s^- centers (phonon-phonon scattering according to Ref. [5]). Curve *b*: Temperature dependence of two phonon decay. Curve *c*: Theoretical fit for D_s^- line (phonon-phonon scattering plus two phonon decay) (after Ref. [51])

where $\bar{n}_i \approx \bar{n}_j = \bar{n}$ was put. The variation of $2\bar{n}+1$ with temperature is shown in Fig. 8 curve (b). Because of Eq. (21) the H_s^- oscillator can decay at most in a four phonon process by a term $V_4(L,i,j,k)a_L a_i^+ a_j^+ a_k^+$ (see Eq. (17), and diagram 4 in Table 6). For the temperature dependence of this process one obtains in analogy $3\bar{n}^2 + 3\bar{n} + 1$. The absence of two phonon decay for the H_s^- oscillator explains the larger lifetime and thereby the smaller halfwidth of the H_s^- line with respect to the D_s^- line at low temperatures. Near 50 K a crossover of H_s^- and D_s^- curves occurs Above this temperature an elastic four phonon scattering process which was proposed by Elliott et al. [5] takes over (diagrams 2 and 5 in Table 6). This process gives a T^2 dependence above the Debye temperature and a T^7 dependence at very low temperatures. The best fit to the KCl : H_s^- data is obtained by using an effective Debye temperature of about 120 K [curve (a) in Fig. 8]. Furthermore, the scattering process is proportional to the square of the amplitude of the localized mode oscillator and thereby explains the factor two in the high temperature halfwidths for H_s^- and D_s^- lines. The theoretical curve (c) for the D_s^- line is obtained

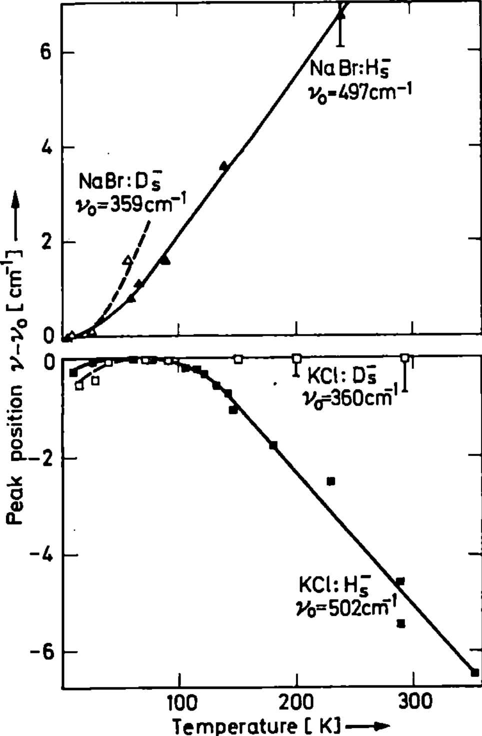

Fig. 9. Temperature dependent frequency shift of the main line (after Ref. [37])

by superimposing the two-phonon decay and the scattering contributions. This is done by adjusting the two phonon decay curve to the saturation halfwidth and adding curve (a) after multiplying it by a factor of 0.5.

The temperature dependent *frequency shift* of the main line was explained by Bilz et al. [51] as being due to the direct anharmonic damping and the thermal expansion of the lattice. The main contributions to the direct anharmonic damping are described by diagrams 1, 3, 6 and 7. Multiphonon processes of this type shift the harmonic localized mode frequency Ω_L to

$$\omega'_L = (\Omega_L^2 + 2\Omega_L \Delta_L(\omega))^{1/2}$$

where $\Delta_L(\omega)$ can be positive and negative as well. The thermal expansion of the lattice is caused by the anharmonicity of *all* band modes. It leads to a softening of the springs and thereby to a decrease in frequency. This additional contribution to the shift must be added to the above equation. It can be estimated from experimental results with mixed crystals [see Section 3.1.6, Eq. (26)] or from uniaxial stress experiments (see Section 3.1.7). The sum of contributions from direct anharmonic damping and thermal expansion may result in a positive or a negative frequency shift as observed for NaBr and KCl with U centers, respectively (Fig. 9).

3.1.4. Sidebands

The origin of the one phonon sideband spectrum is a three phonon process (see diagram 1 in Table 6). The localized oscillator is first *virtually* excited with light by a first order dipole moment. This intermediate state then decays into a final state in which a localized mode quantum is created and simultaneously a lattice phonon either created (sum process through terms $V_3(L, L, i)\, a_L a_L^+ a_i^+$) or annihilated (difference process through terms $V_3(L, L, i) a_L a_L^+ a_i$). The sideband peaks are located at an energy difference ω_i from the main line ω_L where

$$\omega_{\text{Light}} = \omega_L \pm \omega_i .$$

The difference process freezes out for $T \to 0$ because then $\bar{n}_i \to 0$. This can be seen from Fig. 5. The imaginary part of the susceptibility in the sideband region at $T = 0$ K may be written in the density approximation [51]

$$\chi''(\omega) = 16\pi^2 \left(\frac{3 M_1 V_3}{\omega_L \omega_i (1 + \omega_i/2\omega_L)} + M_2 \right)^2 \frac{\varrho(\omega_i)}{\omega_i \omega_L} \tag{22}$$

where $\varrho(\omega_i)$ is the one phonon density for the host crystal, and M_i are expansion coefficients for the dipole moment M. As mentioned in Section 3.1.2, M_2 can be neglected for the U center localized mode in

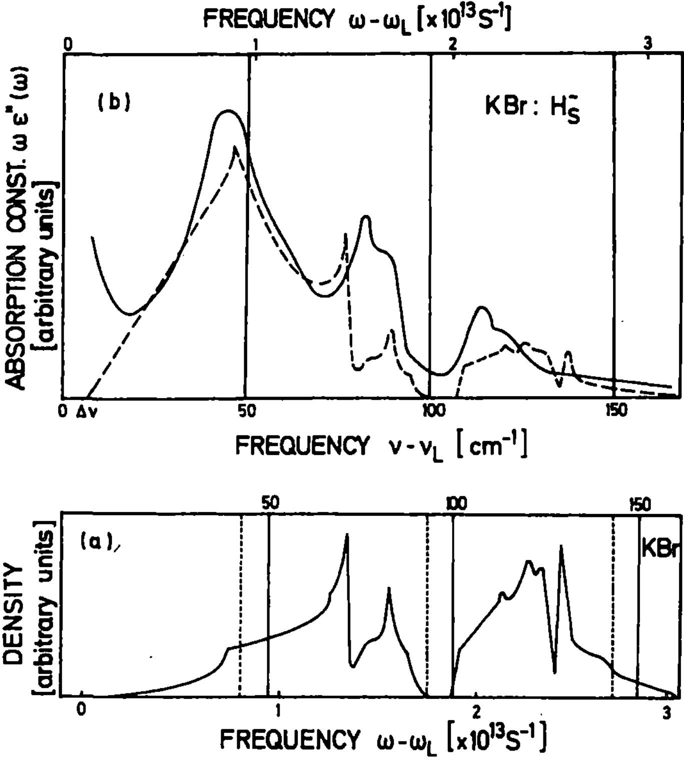

Fig. 10. (a) Phonon-density of states for KBr according to Cowley [83]. (b) Sideband absorption of the H_s^- center localized mode in KBr. Solid line: experimental curve by Fritz et al. [37]; Dashed curve: simple density approximation (see text)

alkali halides. A fit to the experimentally determined sideband spectrum using Eq. (22) is shown in Fig. 10. For $\varrho(\omega_i)$ the unperturbed one phonon density calculated by Cowley [83] was used.

Equation (22), however, is only an approximation, since out of all the possible phonon states, only those of special symmetry can couple to the localized oscillator. This can be verified by using the results of Section 2.2. There it was found that one has to consider only the initial state and the final state in order to find the selection rules. The initial state is that of the photons which transform like T_{1u} in O_h symmetry. The final state is described by $T_{1u} \times \Gamma$, where T_{1u} refers to the first excited state of the localized oscillator. In analogy to Eq. (5) one then finds that the only phonons which can couple to the localized oscillator are contained in the symmetric product

$$(T_{1u} \times T_{1u})_s = A_{1g} + E_g + T_{2g}. \tag{23}$$

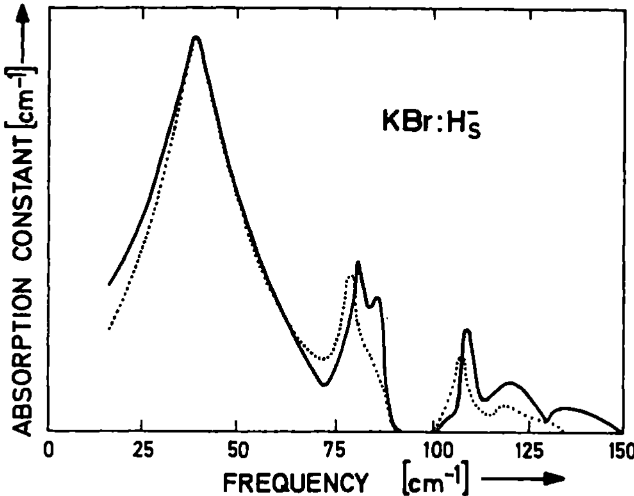

Fig. 11. High frequency sideband structure of the H_s^- center localized mode in KBr. Full line: theoretical curve (after Ref. [22]). Dotted line: modified experimental curve (after Ref. [60])

Their relative contribution to the sideband spectra depends on their coupling strength $A(\Gamma)$; i.e. instead of one "average" coupling parameter as in Eq. (22), three coupling parameters enter into the calculation.

By using these group theoretical aspects, Kühner and Wagner [22] calculated the sideband spectrum by employing the force constant model of Fig. 4. Their result is shown in Fig. 11 by the full line. It is a superposition of the imaginary parts of the perturbed projected Green functions $\hat{G}(\Gamma)$ (see Section 2.4) multiplied by $\omega^{-2}|A(\Gamma)|^2$. This line fits the modified experimentally determined curve (dotted line) much better than the simple density approximation according to Eq. (22). The symmetry vectors of 1 nn coordinates, which can couple to the localized mode are shown, among others, in Fig. 2. In the approximation of Ref. [22] the assumption is made that

$$A(A_{1g}) : A(E_g) : A(T_{2g}) = 1 : 1 : 1 .\tag{24}$$

Experimentally the anharmonic coupling coefficients can be determined from uniaxial stress experiments (see Section 3.1.5). For $KBr : H_s^-$ it was found that the coupling of the localized mode to A_{1g}-type distortions is about a factor of three larger than the coupling to distortions of E_g and T_{2g} symmetry, i.e., the assumption of Eq. (24) is somewhat unrealistic. The local force constants were determined from the main line frequency and a best fit approximation of the sidebands. From this $A01 = -0.50$

Table 7. Local force constant changes $A01$, and $A14$ for H_s^- centers in alkali halides

Substance	$A01$	$A14$
NaF	−0.15	0
NaCl	−0.45	−0.18
NaBr	−0.44	−0.23
KCl	−0.43	0
KBr	−0.42	−0.2
KI	−0.48	−0.28

Force constant changes obtained from a shell model defect (see Fig. 4) calculation fitted to the H_s^- center localized mode frequency and to the sideband shape (after Ref. [46]).

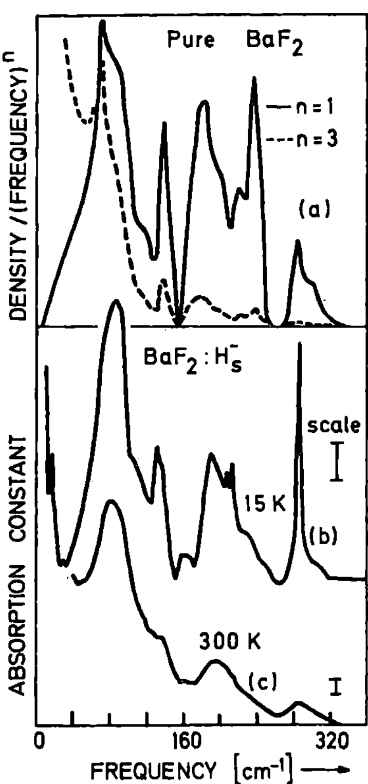

Fig. 12. (a) Phonon-density of states for BaF_2 as determined from neutron scattering at 300 K [84] weighted by ω_i^{-1} (solid line) and ω_i^{-3} (dashed line). (b) High frequency sideband structure of the localized mode fundamental in BaF_2 : H_s^- $(6 \cdot 10^{18} \text{cm}^{-3})$ at 15 K as observed by infrared absorption. (c) Same as (b) but at a temperature of 300 K. (after Ref. [66])

and $A14 = -0.13$ was obtained. Force constant changes $A01$ and $A14$ as obtained in a similar way from a shell model defect calculation by MacPherson and Timusk [46], are listed in Table 7. However, the curvature of the sideband structure does not react very sensitively to changes in $A14$. A much more accurate value of $A14$ is obtained by fitting the acoustic resonance frequency (see Chap. 4).

In spite of oversimplifications, like the assumption of Eq. (24) the remaining deviations between experimental and theoretical curves might reveal additional complications in the interpretation of the sideband structure. In this connection it was shown by Boese and Wagner [63] that the optical Jahn-Teller effect may affect the structural form of the sidebands, particularly in the low frequency region.

In alkaline earth fluorides the sideband structure of the hydrogen localized mode is not yet as well analyzed as for the case of alkali halides. The lattice modes which can couple to the localized mode, and which are observed in the sideband spectra, are of A_1, E, and T_2 symmetry. This is easily seen from Eq. (23) when employed for T_d symmetry. A characteristic sideband spectrum is shown in Fig. 12 (b) and (c) for $BaF_2 : H_s^-$. The sharp peak near $285\,cm^{-1}$ is due to simultaneous excitation of the localized mode and a resonant mode associated with the H^- impurity. Fig. 12 (a) shows a simple density approximation. The curves represent the density of states according to Hurrell et al. [84] but are weighted by ω_i^{-1} (full curve), and ω_i^{-3} (dashed curve) respectively. In contrast to the localized mode sidebands in alkali halides it appears that a ω_i^{-1} weighting factor gives better agreement than a factor ω_i^{-3}!

3.1.5. External Fields

The dependence of the potential of the localized oscillator on changes in the positions of neighboring ions can be studied by stress experiments. Such experiments yield information on the anharmonic coupling coefficients, serve for determining the degree of degeneracy of a mode, and enable one, to study vibrational transitions which are forbidden by symmetry and which might become optically active under stress. Similar information can be derived from experiments with electric fields.

Uniaxial stress lifts the threefold degeneracy of the first excited level ($n = 1$) of the localized oscillator. Fig. 13 shows the splitting of the main line in $KCl : H_s^-$ and $KI : H_s^-$ as a function of applied stress for different polarization directions. E is the vector of the incident light, P is the stress. The frequency shift of single line components is linear with stress. No changes in the band shape and the integrated absorption were found.

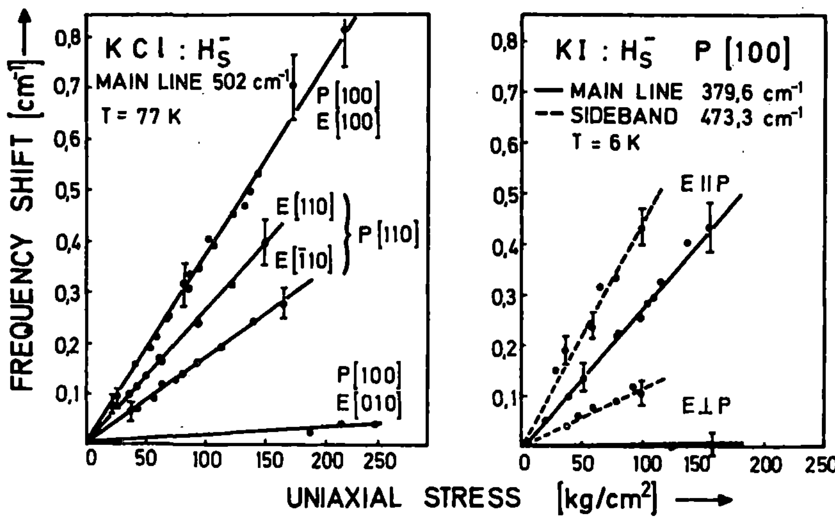

Fig. 13. Splitting of H_s^- center localized mode under uniaxial stress (after Ref. [43])

The strain Hamiltonian for a cubic crystal under uniaxial stress may be written in the form

$$
\begin{aligned}
H = P\{&\alpha(A_{1g})\,(s_{11}+2s_{12})\,r^2 \\
&+ \beta(E_g)\,(s_{11}-s_{12})\,[(2n^2-l^2-m^2)\,(2z^2-x^2-y^2) \\
&+ 3(l^2-m^2)\,(x^2-y^2)] \\
&+ \gamma(T_{2g})\,s_{44}(lmxy+lnxz+mnyz)\}
\end{aligned}
\tag{25}
$$

for an arbitrary direction l,m,n of P relative to crystal axes. s_{ij} are *local* elastic stiffness constants [85]. The stress coefficients α,β,γ play the role of anharmonic coupling constants.[3] In O_h symmetry the strains associated with the coefficients α,β,γ transform like A_{1g}, E_g, and T_{2g}, respectively. In Eq. (25) only terms linear in the displacements of the neighbors and quadratic in the displacement of the H_s^- ion are included.

Stress in [110] direction lifts the degeneracy of the localized mode completely. The third component, which is polarized along [100] has not been measured for KCl : H_s^- (see Fig. 13). The strain energies as calculated from Eq. (25) are given in Table 8 for different polarization directions. From measured frequency splittings the stress coefficients α,β,γ can be calculated. They are listed for various alkali halides and for CaF_2 in Table 9. It is apparent from inspection of the stress coefficients

<hr>

[3] Equation (25) was originally formulated by Gebhardt and Maier [28] in connection with the stress splitting of the electronic F band in alkali halides. There, the coefficients α,β,γ play the role of electron-phonon coupling constants in the optical transition.

Table 8. Energy shift per unit stress for different polarizations of stress P and electric vector of light E

Stress	Electric vector of light	Energy shift/stress
P	E	$-\Delta v/\Delta P$
hydrostatic	arbitrary	$3\alpha(s_{11}+2s_{12})$
001	001	$\alpha(s_{11}+2s_{12})+4\beta(s_{11}-s_{12})$
001	011	$\alpha(s_{11}+2s_{12})+\beta(s_{11}-s_{12})$
001	010	$\alpha(s_{11}+2s_{12})-2\beta(s_{11}-s_{12})$
110	110	$\alpha(s_{11}+2s_{12})+\beta(s_{11}-s_{12})+\frac{1}{2}\gamma s_{44}$
110	100	$\alpha(s_{11}+2s_{12})+\beta(s_{11}-s_{12})$
110	$1\bar{1}0$	$\alpha(s_{11}+2s_{12})+\beta(s_{11}-s_{12})-\frac{1}{2}\gamma s_{44}$
111	111	$\alpha(s_{11}+2s_{12})+\frac{2}{3}\gamma s_{44}$
111	$1\bar{1}0$	$\alpha(s_{11}+2s_{12})-\frac{2}{3}\gamma s_{44}$

Table 9. Anharmonic stress coefficients for high frequency localized modes due to substitutional hydrogen centers

Substance	$\alpha(A_{1g})$ $\times 10^4$ [erg/cm^2]	[cm^{-1}]	$\beta(E_g)$ $\times 10^4$ [erg/cm^2]	[cm^{-1}]	$\gamma(T_{2g})$ $\times 10^4$ [erg/cm^2]	[cm^{-1}]	References
KCl	$2,5\pm10\%$	840 (705)[b]	$0,8\pm10\%$	269 (165)	$0,35\pm10\%$	118 (118)	[43]
KBr	$2,0\pm10\%$	755 (565)	$0,64\pm10\%$	241 (151)	$0,58\pm10\%$	219 (219)	[72]
KI	$0,7\pm20\%$	308 (198)	$0,29\pm20\%$	128 (75)	$0,51\pm20\%$	224 (224)	[43]
CaF$_2$[a] (fundamental)	$5,2\pm4\%$	905	$0,6\pm30\%$	104	$6,3\pm3\%$	1095	[82, 78]
KI (F center in-gap mode)		(125 ± 40)		(35 ± 15)			[86, 87]

[a] In alkaline earth fluorides where the U center has T_d site symmetry the stress coefficients transform like $A_1, E,$ and T_2 respectively.

[b] Values in parantheses are corrected by Nardelli's formula.

that, in alkali halides, the coupling of the U center localized mode to A_{1g} distortions dominates. This shows the inadequacy of the assumption in Eq. (24). For CaF$_2$ the coupling to dilating A_1 and shearing T_2 distortions is comparable in size, and an order of magnitude larger than the coupling to distortions of E symmetry. Uniaxial stress experiments at the D_s^- line in CaF_2 were done by Hayes and MacDonald [78]. The corresponding stress coefficients are by about a factor $\sqrt{2}$ smaller than those for the H_s^- line. This again confirms that the H_s^- and D_s^- vibrations are highly localized. It will be seen in Section 3.2. that the U_1 center localized mode preferentially couples to T_2 lattice modes.

A correlation between the stress coefficients and the anharmonic coefficients of Eq. (13) is not straight forward, and a reliable derivation of such a relation is still missing. However, in alkali halides and alkaline earth fluorides, the approximate magnitude of the stress coefficients compares fairly well with estimates of the third order anharmonic coefficients (see Eq. (17)) as derived from the low-temperature linewidth of the D_s^- line or from the integrated sideband intensity [37, 5, 78].

The investigation of the behaviour of a sideband under stress was done for the 94 cm^{-1} in-gap mode in $KI : H_s^-$ (see Fig. 13) [43]. From the absence of a splitting of this band into more than two components, and from the fact that the relative frequency shift against the main line is practically independent of the polarization and the stress axis, one can conclude that this line is due to an A_{1g} vibration. This is in agreement with a calculation by Gethins et al. [23] which predicts an in-gap mode of A_{1g} symmetry as a consequence of the relaxed force constants pertinent to nearest neighbors of the U center.

In CaF_2, where H_s^- centers have T_d symmetry, stress induced transitions were observed [see Fig. 6(b)] [82]. This is due to the shear strain induced admixture of T_2 states with the infrared inactive levels of the excited $n = 2$ state. In alkali halides a similar activation of second harmonic levels should be possible under the influence of an external electric field.

The behaviour of the H_s^- center localized mode under an applied *electric field* was studied in CaF_2 [78, 82]. In this crystal, the electric field produces a strain of T_2 symmetry; and, as a consequence, a Stark splitting of the excited levels which is linear in the electric field strength was observed. It was found that fields of 10^5 V/cm, when applied in [111] direction, produce a splitting of at most 0.3 cm^{-1}.

3.1.6. Mixed Crystals

The localized mode absorption by U centers in alkali halide mixed crystals was investigated by Mirlin and Reshina [35] and by Barth and Fritz [42]. Fig. 14 shows the main features of the absorption for the example of $KCl : Rb$ with H_s^- centers. Besides the strongest line, ν_0, which has nearly the same position as in pure KCl with H_s^- centers, new lines $\nu_\alpha \ldots \nu_\varepsilon$ occur. On substituting H_s^- centers by D_s^- centers, the

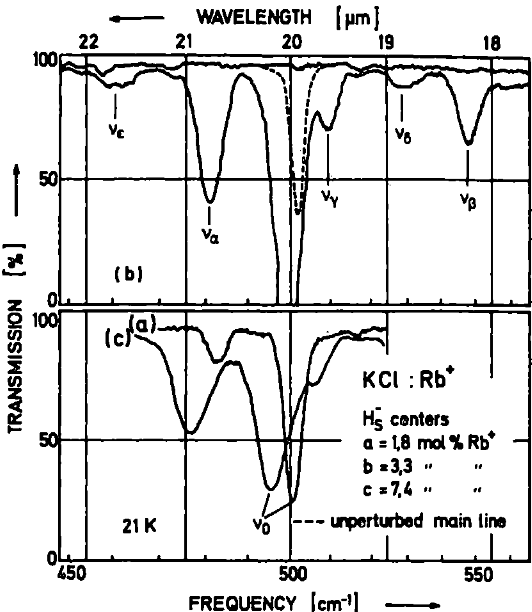

Fig. 14. Infrared absorption spectrum of H_s^- centers in KCl: Rb$^+$, at 21 K. v_0 is the nearly unperturbed H_s^- main line frequency, $v_\alpha \ldots v_\varepsilon$ are new bands which are not seen in pure KCl (after Ref. [42]

additional lines undergo the same isotope shift as the main line, namely $v(H_s^-)/v(D_s^-) = 1.39 \approx \sqrt{2}$. In Ref. [42] lines v_α and v_β were interpreted as being due to U center configurations in which one of the 1 nn is replaced by the Rb$^+$ ion (see Fig. 32 in Section 4.2.2). The line v_γ is supposed to arise from a perturbation due to cations in the third shell. Lines v_δ and v_ε were correlated with U centers in the C_{2v} configuration [see case (d) in Fig. 32]. However, a further study of the relative intensities of lines seems to be necessary. They then could be compared, in analogy to Section (4.2.2), with relative frequencies in different configurations by taking into account the polarization and the degree of degeneracy in different modes.

Besides the frequency splitting the doping also causes a shift in the main line v_0. This is shown in Fig. 15. The decrease in frequency with increasing Rb$^+$ doping can be understood from an overall expansion of the lattice caused by the Rb$^+$ ions. Such an expansion of the lattice was verified from X-ray data by Gnaedinger [100]. The frequency shift can be compared with the results obtained from the uniaxial stress experiments reported in Section 3.1.5. From Table 8 one can easily derive the relation

$$\Delta v = (3\Delta a/a)\,\alpha(A_{1g}). \tag{26}$$

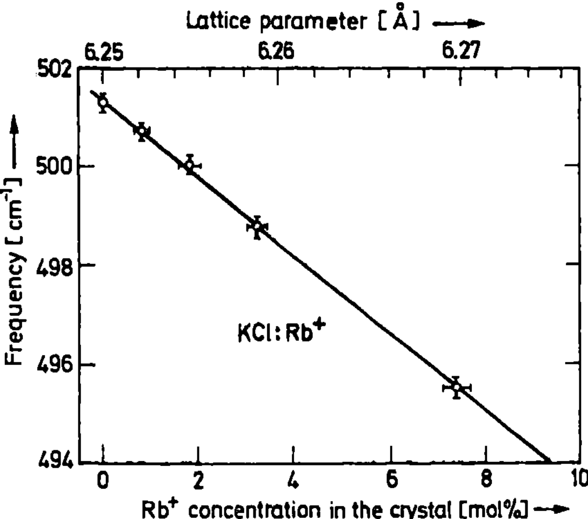

Fig. 15. Line positions ν_0 for KCl:Rb⁺ with H_s^- centers versus doping concentration at 21 K. The lattic parameter is interpolated using a Vegard relation known to hold at room temperature. The bands $\nu_a \ldots \nu_y$ show the same dependence (after Ref. [42])

From Fig. 15 one obtains $\alpha(A_{1g}) \approx 610\,\mathrm{cm}^{-1}$. This value agrees fairly well with the value tabulated for KCl with H_s^- centers in Table 9, when one takes into account the fact that the lattice parameter change in the mixed crystal is about one order of magnitude larger than in the stress experiments.

3.1.7. Pair Centers

The near infrared absorption spectra of $H_s^- H_s^-$, $D_s^- D_s^-$, and $H_s^- D_s^-$ pair centers in KCl were studied by de Souza et al. [47]. For the aligned [110] pair centers in C_{2v} symmetry six non-degenerate localized modes are expected: three in phase and three out of phase modes which are polarized in [110], [1$\bar{1}$0] and in [001] directions. The out of phase modes are infrared inactive for $H_s^- H_s^-$ and $D_s^- D_s^-$ pairs, but active for the mixed $H_s^- D_s^-$ centers. The infrared active in phase modes will be polarized in the longitudinal (L), and in the transverse (T_1 and T_2) directions of the pair. This is shown in the insert of Fig. 16. Using polarized light, the infrared measurements of the localized mode spectrum of crystals containing [110] aligned pairs revealed the position and polarization of the pair modes. The spectra are shown for the case of $H_s^- H_s^-$ pair centers in Fig. 16. Three lines at 463.5, 512.5, and 535 cm⁻¹ are found

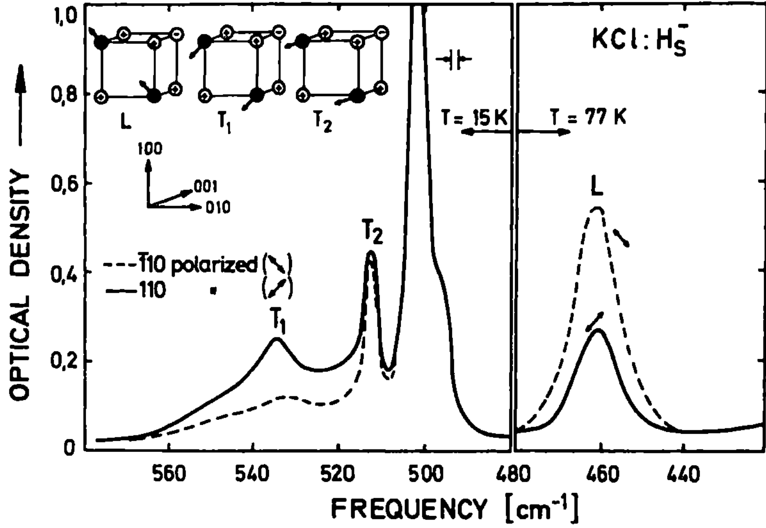

Fig. 16. Absorption spectrum of KCl : H_s^- ($3 \cdot 10^{18}$ cm^{-3}) after production of aligned $H_s^-\,H_s^-$ pairs (with preferential [$\bar{1}10$] orientation as indicated in the upper left-hand corner), measured for [$\bar{1}10$] and [110] polarized light incident in a [001] direction. The phonon sideband of the H_s^- localized mode at 565 cm^{-1} has been subtracted. The band at 463 cm^{-1} was measured at 77 K in order to avoid errors from insufficient resolution (after Ref. [47])

Table 10. Measured and calculated line positions for pair centers (after Ref. [88])

		$H_s^-\,H_s^-$		$D_s^-\,D_s^-$		$H_s^-\,D_s^-$	
		exp.	calc.	exp.	calc.	exp.	calc.
[110]	in phase	463	463	331	329	351	347
	out of phase		537		380	508	507
[1$\bar{1}$0]	in phase	535	535	375	379		348
	out of phase		465		330	511	506
[001]	in phase	512	512	368	363		353
	out of phase		492		349		503

which can be correlated with L, T_2 and T_1 modes, respectively. The measured line positions for $H_s^-\,H_s^-$, $D_s^-\,D_s^-$ and $H_s^-\,D_s^-$ centers are listed in Table 10 together with calculated values which are based on a linear chain model.

3.2. U_1 Centers (H_i^-, D_i^- Centers)

3.2.1. Perturbed U_1 Centers

In analogy to U centers, U_1 centers give rise to localized mode absorption in the near infrared. However, in alkali halides, the interaction with nearby anion vacancies (α centers) – which are simultaneously produced in the $U - \alpha$ process (see Section 2.1) – results in a point symmetry of U_1 centers which is in general lower than cubic (T_d).

The near infrared vibrational spectrum, which was first investigated by Fritz [69], essentially consists of three groups of lines (see Fig. 17). Single groups of lines can be correlated with H_i^- centers of different thermal stability. The interaction between U_1 centers and α centers is of elastic and electric (Coulomb) nature. Its origin is the charge of the defects. This interaction results in the tendency for thermal recombination of U_1 centers with α centers to U centers (see Fig. 18), and in the splitting of the absorption lines. Both, the thermal instability of U_1 centers, and the amount of splitting in the vibrational spectrum, are inversely proportional to the relative distance between the defects.

It is not completely understood why essentially three distinct pairs of vacancy – interstitial ion configurations exist. From the observation of three line groups and three relative sharp annealing steps one can conclude that the interstitial ions occupy in each stage a single site or

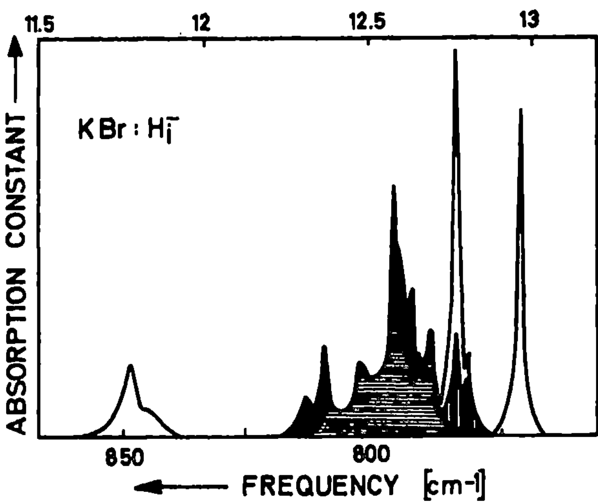

Fig. 17. Infrared absorption spectrum due to H_i^- ions in KBr after low-temperature (20 K) irradiation in the ultraviolet U band. The various line groups disappear after warming to the following temperature regions: white, $90-110$ K; vertical striping, $130-160$ K; horizontal striping, $190-230$ K (after Ref. [69])

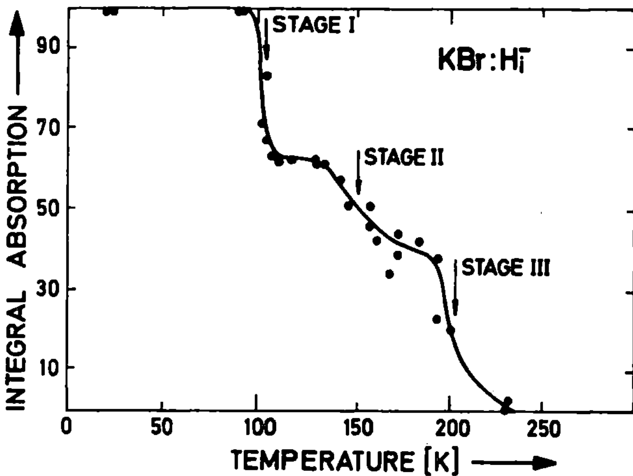

Fig. 18. Decrease of the total integral absorption in the spectrum of Fig. 17 as a function of annealing temperature (after Ref. [69])

a small number of almost equivalent sites. Notice that the U_1 centers undergoing annealing between 190 and 230 K (stage III) still feel the influence of the vacancy. However, it was shown recently for the case of KBr and KI that a careful thermal annealing procedure leads to H_i^-, and D_i^- centers, whose point symmetry is *not* lowered by short-range interaction with nearby anion vacancies. The symmetry of these "free" H_i^-, D_i^- centers turns out to be cubic (T_d).

3.2.2. The Main Line

The near infrared vibrational spectra of nearly "free" H_i^- and D_i^- centers in KBr and KI are shown in Figs. 19 and 20. In analogy to the near infrared vibrational absorption of U centers, the spectrum consists of a main line and sidebands. Table 11 summarizes some of the experimental data. Preliminary results on the main line frequencies in LiF, KCl, RbCl, and RbBr are included.

The main line shows a temperature dependent half width which follows a T^2 law above about 80 K. When cooling the sample to liquid helium temperature in KBr: H_i^- and KI: H_i^- the line width becomes $\lesssim 1.5\,\mathrm{cm}^{-1}$ (see Table 11). This remaining half width might still be influenced by a "small" interaction with vacancies. However, no splitting in the main line was observed! This leads to the conclusion that for free U_1 centers the hydrogen moves in a cubic potential of T_d symmetry; i.e.

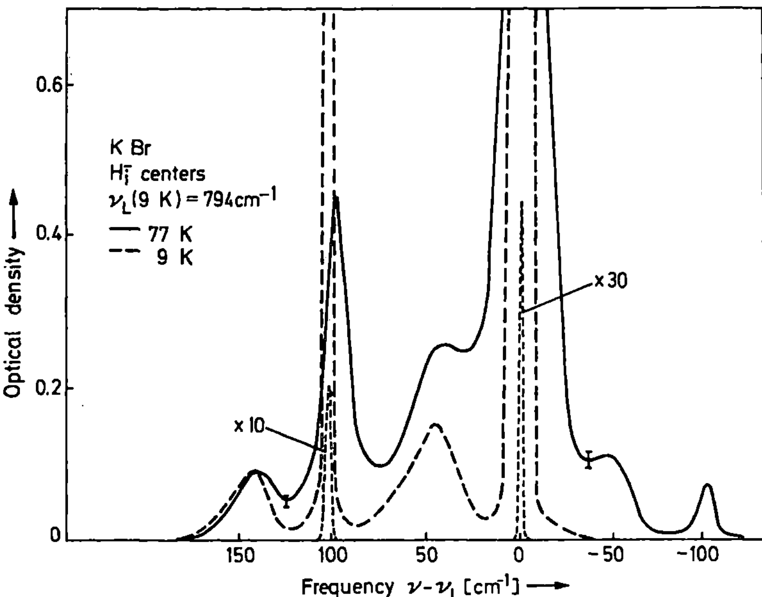

Fig. 19. Analyzed near-infrared spectra of "free" H_i^- centers in KBr at 77 and 9 K (after Ref. [72])

the H^- or D^- ion is sitting in the middle of a double tetrahedron formed by anions and cations, respectively. The vibrational transition then occurs from a non-degenerate A_1 ground state to a three-fold degenerate $T_2(1)$ excited state. A model for this vibration is included in Fig. 41. It is in contrast to the model suggested by Gross and Bron [70]. The strong localization of the U_1 center mode is seen in the isotope shift which is approximately $\sqrt{2}$ for H_i^- and D_i^- centers as expected from the mass ratio (see Table 11 and Sections 2.3 and 3.1.1). An absorption due to second or third harmonic transitions, which are allowed by symmetry (see 3.1.1), was not yet found. From the measurements one can conclude that the intensity of these transitions is $\lesssim 10^{-3}$ when compared with the A_1 to $T_2(1)$ transition.

Interstitial hydrogen ions were found also in CaF_2, but only when the crystals were doped with rare earth ions. Fig. 21 shows the infrared absorption spectra of rapidly quenched hydrogenated CaF_2 crystals containing Gd ions. The spectra show four lines. Two of these lines, one at $965 \, cm^{-1}$ – which was observed also in pure CaF_2 (see Section 3.1) – and one at $1296 \, cm^{-1}$, have frequencies which are independent of the rare earth ions. From a number of experiments (see Ref. [7]) and the fact that no higher harmonics could be observed, the

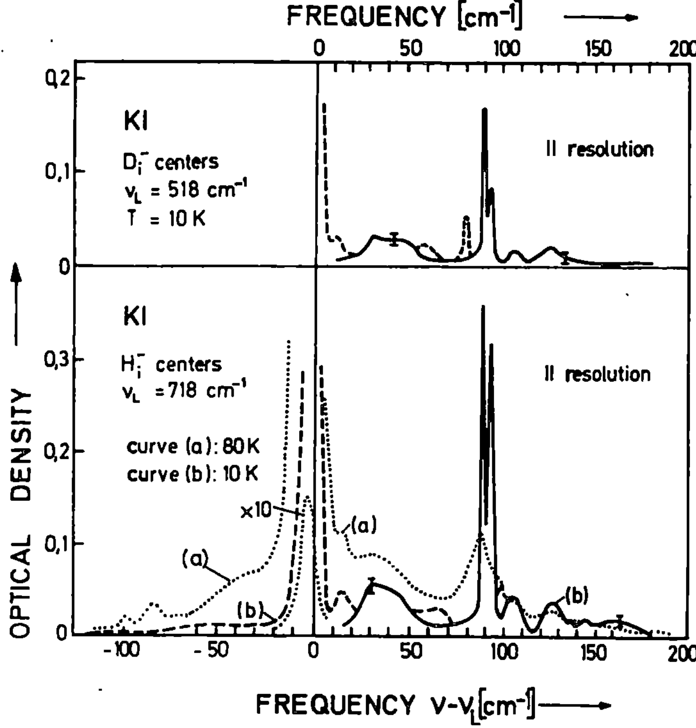

Fig. 20. Near infrared spectra of H_i^- and D_i^- centers after "purification" by annealing. Broken lines: experimental curves. Full lines: analyzed sideband shapes (after Ref. [71])

Table 11. Main data on the high frequency localized modes and in-gap modes due to interstitial hydrogen centers in alkali halides and alkaline earth fluorides

Sub-stance	Localized mode frequency $v_L(H_i^-)$ [cm^{-1}]	Halfwidth $\Delta v_L(H_i^-)$ [cm^{-1}]	Freq. ratio $v_L(H_i^-)/v_L(D_i^-)$	Temperature [K]	Intensity ratio I_{SB}/I_T	In-gap mode frequency v_G [cm^{-1}]	References
LiF	2100			130			[73, 74]
KCl	850			20			[69]
KBr	794	<1.5	1.40	9	0.20	98.7	[69, 72]
KI	718	<0.8	1.39	10	0.15	86.7	[71, 86]
RbCl	800			20			
RbBr	748			55			
CaF$_2$	1310	15		77			
	1296	30	1.39	300			[15]

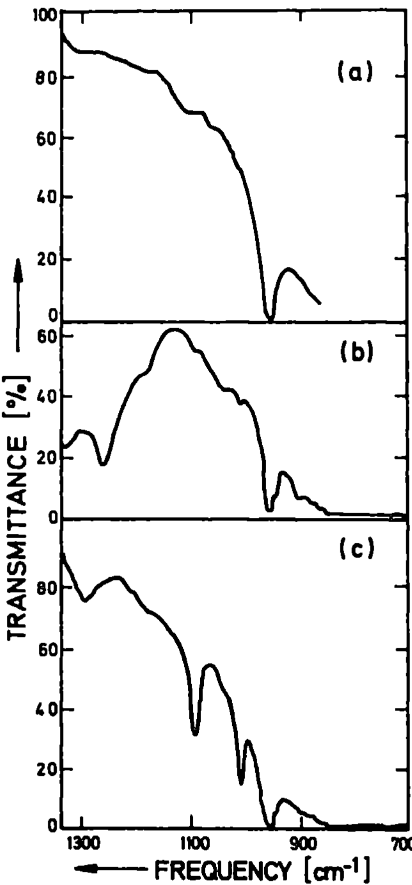

Fig. 21. Infrared absorption spectra of (a): hydrogenated pure CaF_2. (b): slowly annealed (20°/h) hydrogenated CaF_2 containing Gd. (c): rapidly quenched hydrogenated CaF_2 containing Gd (after Ref. [7])

$1296\,\mathrm{cm}^{-1}$ line was correlated with hydrogen ions which are well separated from rare earth ions and which are sitting on an interstitial position at the center of an empty fluorine cell. These hydrogen ions have O_h site symmetry. A model is shown in Fig. 22. In deuterated crystals a similar line occurs near $932\,\mathrm{cm}^{-1}$. The ratio of absorption frequencies is 1.39 (see Table 11). The frequencies of the remaining two lines at $1007\,\mathrm{cm}^{-1}$ and $1093\,\mathrm{cm}^{-1}$ in the spectrum of Fig. 21 vary with the particular rare earth ion present. These lines could be correlated with interstitial hydrogen ions adjacent to rare earth ions which split the first excited T_2 level of the hydrogen oscillator twofold. The site

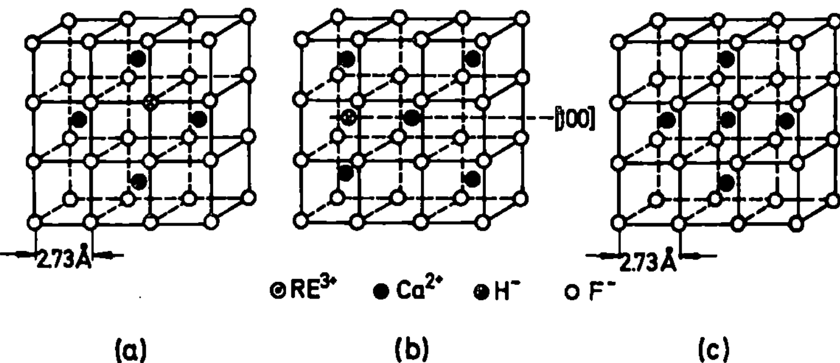

Fig. 22. Structure of U_1 centers in CaF_2. (a): pure CaF_2. (b): the axial-site center. (c): the cubic-site center (after Ref. [7])

symmetry of these hydrogen ions is C_{4v}. It was established from the number of second harmonics (three) observed, and from ESR measurements on crystals containing Ce and Nd ions respectively.

3.2.3. Sidebands

For an identification of the one phonon sidebands, U_1 center concentrations of some $10^{17}/cm^3$ are necessary. With these "high" concentrations it was not possible to completely suppress a small amount of perturbed U_1 centers. However, one can identify the one phonon sideband spectra by using the following criteria:

a) The frequency displacements of sideband peaks should be equal for H_i^- and D_i^- centers.

b) The peaks should be roughly symmetric about the main line; the high and low frequency peaks have temperature dependent intensities according to $\bar{n}+1$ and $\bar{n}$, respectively (see Sections 3.1.2 and 3.1.4).

Figure 19 shows the sideband spectrum for $KBr:H_i^-$ which was identified according to these criteria (see also Ref. [72]). The results for $KI:H_i^-, D_i^-$ are shown in Fig. 20. The broken lines indicate a "rest" of perturbed U_1 centers. The frequencies of the absorption lines arising from perturbed H_i^- and D_i^- centers are in a ratio $v_L(H_i^-)/v_L(D_i^-) \approx \sqrt{2}$. Such a ratio is expected for perturbed main lines (see Section 3.1.6).

The phonon sidebands arise through anharmonic coupling between the localized mode and lattice band modes – in analogy to the U center localized mode sidebands. The low frequency sidebands disappear for $T \to 0$ (see 3.1.4). For the ratio of integral absorption of sidebands and the total absorption it was found experimentally

$$I_{SB}/I_T \approx 0.2 \qquad \text{for KBr}$$

and   (27)

$$I_{SB}/I_T \approx 0.15 \qquad \text{for KI.}$$

This ratio is a measure of the amount of anharmonic coupling [37, 51]. It is roughly the same as for the high frequency localized mode of U centers [37]. This means that the total third order anharmonic coupling is about the same for U and U_1 centers. However, the U_1 center localized mode couples preferentially to one mode, say Γ, whose frequency in the sideband spectrum occurs in the band gap. This can be seen from Figs. 19 and 20. The ratio of integral absorption of this single line and the total sideband absorption is experimentally

$$I_\Gamma/I_{SB} \approx 0.5 \qquad \text{for KBr}$$

and

$$I_\Gamma/I_{SB} \approx 0.2 \qquad \text{for KI}$$ (28)

where I_Γ for KI contains only the low energy component of the doublet.

From Eq. (23), when employed to T_d symmetry, one can easily verify that the only lattice modes which can couple to the localized mode, are of A_1, E, or T_2 symmetry. Among the T_2 modes are polar modes T_2^P as well as rhombic distortions T_2^S (shear modes). The fact that polar T_2^P modes are infrared active would suggest the possibility of a search for these modes by means of far infrared spectroscopy; the results of such investigations are reported in Section 4.6.

3.3. U_2 Centers (H_i^0, D_i^0 Centers)

The only results on interstitial hydrogen atoms in alkali halides were published for LiF [73]. In γ and neutron irradiated LiF : OH$^-$ a line at $2200\,\text{cm}^{-1}$ is observed which was correlated with the localized vibration of H_i^0 centers. It would be nice, however, if one could further check the results, e.g. by starting out from LiF : OD$^-$ material.

Low temperature X-ray or UV irradiation of hydrogenated CaF_2 crystals containing rare-earth ions converts hydrogen ions on interstitial lattice positions into interstitial neutral hydrogen atoms. This was established for centers having C_{4v} symmetry (see Section 3.2.2) by ESR and ENDOR measurements in crystals containing Ce and Nd ions respectively [7]. The existence of a localized mode due to interstitial hydrogen atoms appears from the vibronic spectra. Vibronic transitions separated from their parent lines by $767\,cm^{-1}$ and $565\,cm^{-1}$ were observed for hydrogenated and deuterated crystals, respectively. Direct infrared localized mode absorption of interstitial hydrogen atoms in CaF_2 occurs near $640\,cm^{-1}$ [75]. The effective charge associated with this vibration is about 0.07 electrons (compared to 0.7 to 0.9 electrons for U centers in alkali halides).

3.4. Interstitial Hydrogen Molecules ($H_{2,i}^0$ Centers)

Akhvlediani and Politov [73] observed in γ and neutron irradiated $LiF:OH^-$ crystals a line at $2000\,cm^{-1}$ which they believe as being connected with interstitial hydrogen molecules. A further clarification of the experimental situation seems to be necessary (see Section 3.3).

4. Infrared Vibrational Absorption: In-Gap Modes and Resonant Modes

In-gap modes and acoustic resonant modes of electron and hydrogen centers were studied experimentally only in the alkali halides KBr and KI. Table 12 summarizes the infrared vibrational frequencies. From Table 12 and Table A1 in the Appendix, one finds that $U(H_s^-, D_s^-)$ and F' centers give rise to acoustic resonant mode absorption ([18, 72, 86, 89–91]; [91]), $F, F_A(Na)$, and $U_1(H_i^-, D_i^-)$ centers to in-gap mode absorption ([72, 86, 87, 91]; [86]; [71, 72, 86]). The only theoretical investigations were performed for U centers [89, 92], and for F centers [87, 94, 95].

In analogy to high frequency localized modes, the frequency positions and splitting of in-gap modes and acoustic resonant modes allow conclusions on local force constants and on local site symmetries of a defect.

For the case of U centers and U_1 centers, where high frequency localized mode absorption *and* acoustic resonant mode or in-gap mode absorption is observed, one can try to calculate the far infrared line frequencies and their absorption shape by using only the parameters determined from the near infrared spectra. The comparison of calculated and measured far infrared spectra yields information on the adequacy and the sensitivity of a lattice dynamical model.

For the case of U_1 centers the lack of inversion symmetry allows the direct far infrared excitation of phonons, which can be excited indirectly in the localized mode sideband spectra as well. From the comparison of near and far infrared spectra one can experimentally determine the

Table 12. In-gap mode and resonant mode absorption of electron and hydrogen centers in alkali halides

Substance	Frequency $\nu_{G,R}$ [cm^{-1}]	Half-width $\Delta\nu_{G,R}$ [cm^{-1}]	Temperature [K]	References
KBr				
$: H_s^-$	89 ± 0.5		7	[89, 90]
$: D_s^-$				
$: F$	99.60 ± 0.03			
	99.07 ± 0.04	<0.1	1.2	[87, 90]
	98.50 ± 0.05			
$: H_i^-$	98.7 ± 0.5	<1.8	9	[72]
$: D_i^-$	98.7 ± 0.5	<1.8	9	[72]
KI				
$: H_s^-$	61 ± 0.5			[18, 89, 91]
$: D_s^-$	60.50 ± 0.5			[89]
	60.25 ± 0.5			[71]
$: F$	82.62 ± 0.02			
	81.98 ± 0.02	<0.1	1.2	[91, 87]
	81.19 ± 0.05			
$: F_A(\text{Na})$	(82)	<1.2	4.2	[86]
	80.05			
$: M$	$-$			[86]
$: F'$	68.5 ± 0.5	<1	7	[91]
$: \alpha$	$-$			[86]
$: H_i^-$	86.7 ± 0.5	<1.3	7	[71]
$: D_i^-$	86.7 ± 0.5	<1.3	7	[71]

relative magnitude of anharmonic coupling coefficients for phonons of special symmetries.

For F centers extensive experimental and theoretical studies of the electronic absorption band, which shows strong temperature-dependent broadening due to electron phonon coupling, did allow the prediction of some lattice dynamical properties [1, 93] Vice versa, the direct investigation of the vibrational spectra yields a better understanding of the electronic transition, and of the electron phonon coupling. The fact that the electronic absorption band does not exhibit distinct vibrational structure, but is smooth and nearly Gaussian (strong coupling case), allows one to obtain a reasonable quantitative picture of the band shape by employing oversimplified dynamical models for the lattice – as e.g. the configurational diagrams (see e.g. Ref. [96]). However, only a detailed treatment of the impurity-lattice vibrational problem can reproduce the shape of the electronic F band quantitatively without making use of fitted ad hoc parameters. Such a calculation may, in addition, supply insight into the electron-phonon energy exchanges.

This treatment allows one to determine from the infrared absorption and Raman spectra, local force constants which may serve for a calculation of local elastic relaxations. These relaxations, in turn, affect the overlap between the wave functions of the electron and its neighbors, and must therefore be taken into account in a calculation of the electronic energy of the defect.

Both, electron and hydrogen centers may give rise to defect induced absorption, which yields information on host lattice band mode frequencies and band mode densities.

4.1. U Centers (H_s^-, D_s^- Centers)

Difference spectra of the far infrared absorption of KBr and KI containing U centers are shown in Figs. 23 and 24. The absorption bands are located in the acoustic phonon band of the host lattice. Even at low temperatures their linewidth remains large. The substitution of H_s^- centers by D_s^- centers in KI results only in a shift of the central component of the threefold structured line near 61 cm^{-1}. This central line shifts to

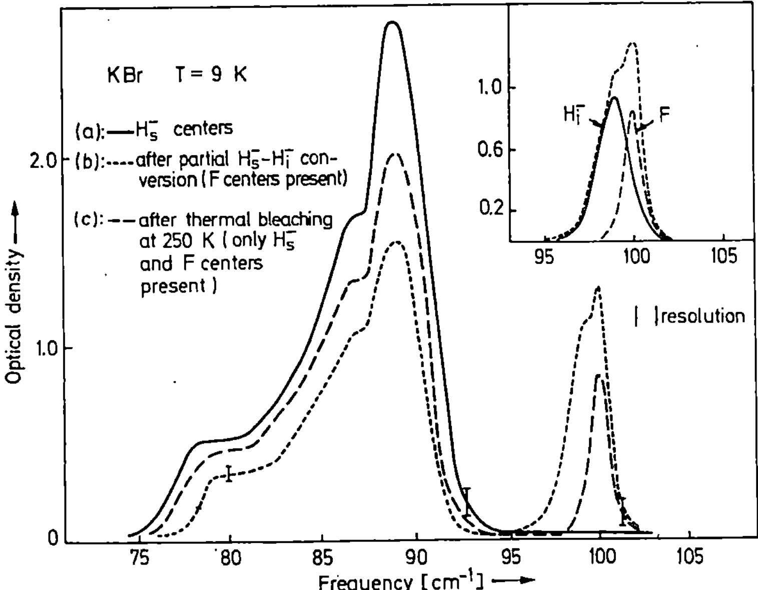

Fig. 23. Acoustic resonant mode absorption of H_s^- centers in KBr (full line). Broken line shows the result of the $U - \alpha$ process: H_s^- centers are converted to H_i^- centers and α centers (a small amount of F centers is simultaneously produced). H_i^- centers and F centers lead to in-gap mode absorption (after Ref. [72])

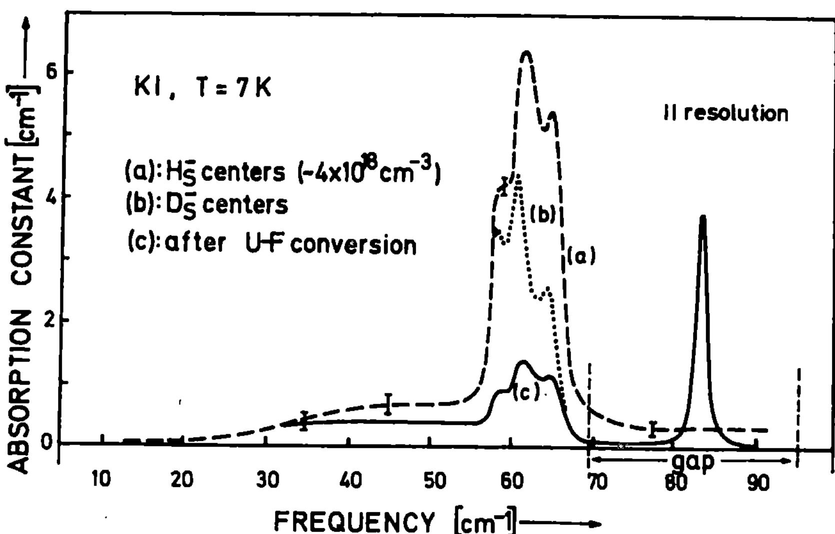

Fig. 24. Acoustic resonant mode absorption of H_s^- centers (broken line) and D_s^- centers (dotted line) in KI. Full line shows the result of the $U-F$ conversion (after Ref. [71])

longer wavelengths by about $0.75 \, \text{cm}^{-1}$. Only this line can be correlated with a $T_{1u}(2)$ vibrational mode of the hydrogen ion (see Figs. 2 and 3). The remaining structure is the defect induced *one* – phonon band mode absorption.

The absorption of originally inactive host lattice band modes yields information on their density of states; e.g. one finds that the maxima and discontinuities occuring in the absorption spectra outside of the resonance do not depend upon the type of defect. As an example, the U center induced acoustic band mode absorption in KBr is compared with the absorption induced by Na^+ and Cl^- centers (see Fig. 25). The frequency positions of discontinuities (labelled by arrows) coincide for vanishing defect center concentrations. They can be correlated *directly* to singularities in the phonon spectrum of the host lattice (see last row in Table 13 and Fig. 10a). The character of the singularity, however, may change. This can be concluded from a comparison with phonon density of states curves as derived from neutron inelastic scattering experiments (see Fig. 10a).

In the T_{1u} (2) resonant mode, the hydrogen moves nearly in phase with 1 nn (see Figs. 2 and 3). The vibrational amplitude is *not* localized at the defect site, in fact, the amplitudes at 1 nn sites and at the defect site are of about the same magnitude. One can approximately describe this vibration by a rigid oscillator – consisting of the defect and 1 nn –

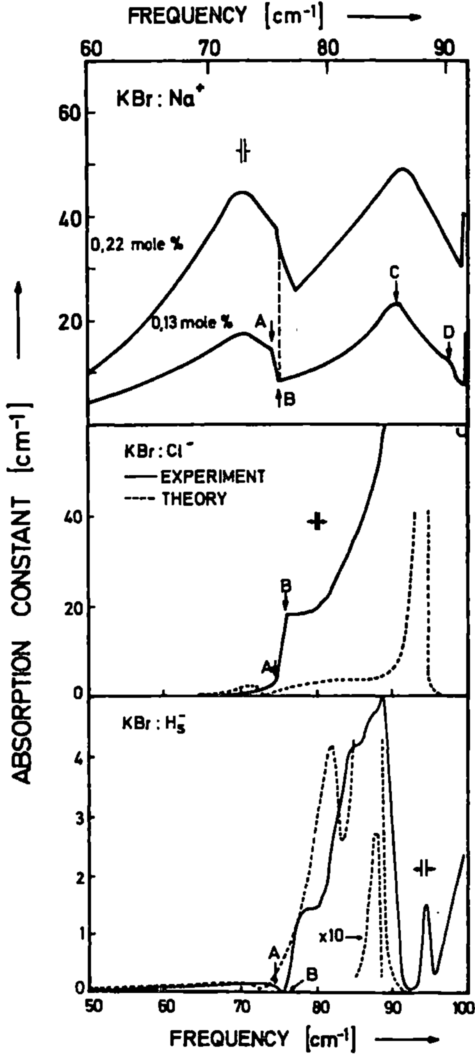

Fig. 25. Acoustic band mode absorption induced by Na^+, Cl^-, and H_s^- centers in KBr (see also Fig. 23). Discontinuities in the spectra are marked by A, B, C, and D. Note that in the case of Cl^- the slope has reversed (after Ref. [98])

vibrating against 4 nn (see Section 2.3). Therefore, the isotope shift expected is small. In $KI : H_s^-$, D_s^- the ratio of the resonance frequencies found experimentally is

$$\nu_R(H_s^-)/\nu_R(D_s^-) \approx 1.01$$

Table 13. Positions of observed singularities in KBr crystals (see Fig. 25) with H_s^-, Cl^-, and Na^+ centers at 11 K

Defect	Mol% defect	A [cm^{-1}]	B [cm^{-1}]	C [cm^{-1}]	D [cm^{-1}]
H_s^-	0.001	75	75.5	85	86
Cl^-	2	74.7	76	–	–
Na^+	0[a]	74.68	75.2	85.22	89.73
Shell model		70.7	71.3	83	88

[a] Extrapolated to zero concentration.

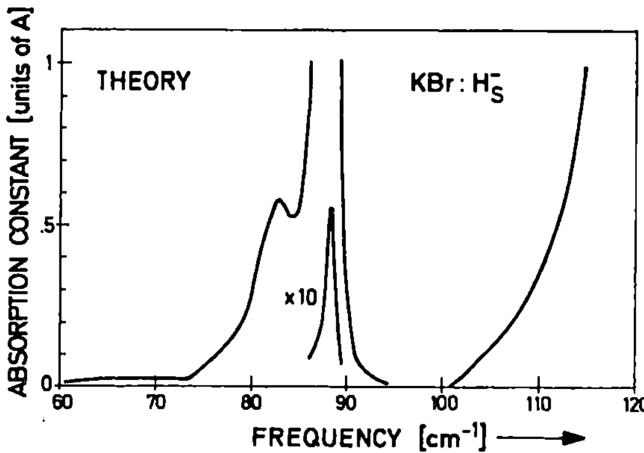

Fig. 26. Calculated absorption spectrum for $KBr : H_s^-$ [The unit A is $4\pi e^2 \, p(n^2(\infty) + 2)^2 / 9n(\omega)\,\mu c \times 10^{-12}$ sec] (after Ref. [89])

A three dimensional model calculation on the far infrared absorption of KBr and KI with U centers was performed by Woll et al. [89]. These authors employ the same force constant model, which they earlier used for a theoretical fit of the high frequency localized mode absorption (see Section 3.1). This model enables one to predict the far infrared absorption spectra without introducing any further parameters. The results are shown in Figs. 26 and 27. The comparison with the experimental curves in Figs. 23 and 24 shows that the overall shape of the spectrum is reproduced fairly well. The very sharp peak in both of the theoretical curves is the resonance. The fact that these peaks are narrower than the experimental ones might be due to the neglect of anharmonic contributions. Furthermore, discrepancies occur in the exact resonance positions, especially for KI. These discrepancies could be due either to faults in the neutron-derived shell-model density of states or to inadequacies in the

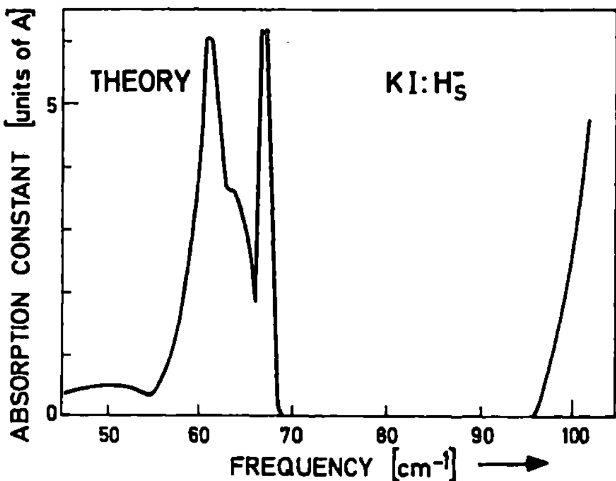

Fig. 27. Calculated absorption spectrum for $KI:H_s^-$ (see caption of Fig. 26 for unit A.) Model evidently predicts the resonance too high by $6-10\,cm^{-1}$ (after Ref. [89])

force constant model for the defect. From the results in Section 4.2, it is believed that the latter is mainly responsible for the discrepancies. The far infrared mode frequencies are extremely sensitive to force constant changes A14 – much more than the sideband shape. A shell model fit of the force constant changes A01 and A14 to the main line frequency of the localized mode and the resonant mode frequency, respectively, yields

$$A01 = -0.43; \quad A14 = -0.20 \text{ for KBr}$$

and

$$A01 = -0.48; \quad A14 = -0.23 \text{ for KI}.$$

In spite of the reservations which should be made when one compares local force constants (see Section 4.2.4), the deviation in $A14$ from the value which was obtained by Kühner and Wagner by fitting the shape of the localized mode sideband for $KBr:H_s^-$ ($A14 = -0.13$) is obvious.

4.2. F Centers

4.2.1. In-Gap Mode and Acoustic Band Mode Absorption

According to the simple model of Section 2.3 one would expect for F centers a vibrational mode of the type $T_{1u}(2)$. The frequency of this mode should lie in the acoustic optical band gap. In this vibration the electron is "carried along" by 1 nn. In fact, for F centers this vibration was found experimentally. The results of the $U \to F$ conversion in KBr and KI are included in Figs. 23 and 24 respectively.

These spectra were obtained by comparing the transmission of "doped" and "undoped" samples. In KBr the F center mode is found at the upper gap-edge near 100 cm^{-1}, while a similar line in KI occurs almost in the middle of the gap near 83 cm^{-1}. In both cases the linewidth is determined by the resolution of the spectrometer which is about 1 cm^{-1} in these experiments.

A detailed theoretical investigation of the lattice dynamics of the F center in connection with the band shape of the electronic absorption, the F center induced first-order Raman spectra (see Chap. 5), and the infrared absorption spectra was performed by Benedek and Mulazzi (1969) [94]. The calculation of these authors is based on one of Hardy's deformation dipole models. Two different dynamical force constant models are suggested. The infrared absorption spectra calculated from

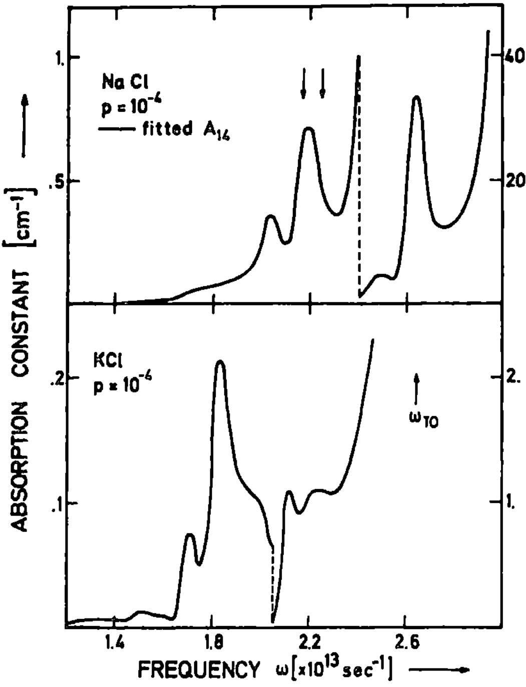

Fig. 28. Infrared absorption spectra due to F centers in NaCl and KCl as obtained for the extended dynamical model. The arrows indicate the frequencies for which the resonant denominator is zero. Note that the vertical scale is changed in the highfrequency region near the reststrahl frequency ω_{TO} (after Ref. [94])

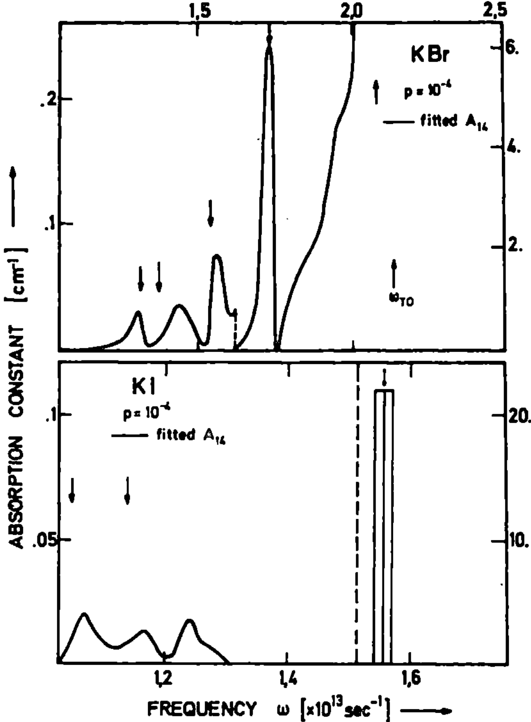

Fig. 29. Same as Fig. 28, but for KBr and KI. In both spectra the induced T_{1u} in-gap modes and their strengths are evident (after Ref. [94])

one of their models, the one which corresponds to the model of Fig. 4, are shown in Figs. 28 and 29. The force constants A_{01} and A_{14} are fitted to the experimental values for the halfwidth of the electronic F band and the Huang – Rhys parameter for zero temperature. The concentration of F centers was assumed to be $p = 10^{-4}$ (see Section 2.4). All calculated spectra show defect induced band mode absorption. Zeros in the resonance denominator (see Section 2.4) are indicated by arrows. In NaCl they appear in the region between the acoustical and transverse optical modes, where the host lattice phonon density turns out to be large. In this case the absorption shape therefore essentially exhibits the T_{1u} projected density rather than resonant mode structure. The band mode absorption due to F centers was not yet verified experimentally.

The spectra for KBr and KI show strong in-gap modes (at 90 cm^{-1} and 82 cm^{-1} respectively) whose frequencies are in agreement with the experimentally determined frequencies within 10%. The remaining deviations in the exact line positions are mainly due to the fact that the force constant fit mentioned above is not very sensitive to changes in A_{14}. A very sensitive method for obtaining local force constants A_{01} and A_{14} from the infrared data, independent from data on the electronic transition, is reported in Section 4.2.2.

A calculation on the F center induced in-gap mode absorption in NaBr, NaI, KBr and KI was performed by Singh and Mitra [95]. These authors use a model, earlier suggested by Jaswal [97], in which the ions are assumed as being rigid, and where the impurity and its six 1 nn are treated as a vibrating molecule.

The force constant – in analogy to A_{01} – was fitted to the frequency of the electronic F band, thereby employing the picture that the electronic absorption band is in reality the local mode of the F center electron (see beginning of Chap. 3). The calculated frequency position of the $T_{1u}(2)$ in-gap mode for KBr and KI is $92.6\,\mathrm{cm}^{-1}$ and $81.2\,\mathrm{cm}^{-1}$, respectively. In-gap modes in NaBr and NaI should occur near $120\,\mathrm{cm}^{-1}$ and $103\,\mathrm{cm}^{-1}$, respectively.

4.2.2. The Isotope Splitting

Figs. 30 and 31 show absorption spectra of the F center in-gap mode for KBr and KI with a resolution of about $0.1\,\mathrm{cm}^{-1}$. In both cases a threefold structure with lines A, B, and C is revealed (with respect to the line near $80\,\mathrm{cm}^{-1}$ in KI see Section 4.3). These lines are extremely sharp. Their real half width is not yet resolved. The relative intensities of these lines are constant and independent of the method of F center production. No frequency shift in lines A and B was observed on changing the F center concentration between 10^{17} and $3.10^{18}\,\mathrm{cm}^{-3}$.

The ratio of the integral absorption, $\int\alpha(v)\,dv$, of lines A, B, and C is experimentally for the case of KI

$$I(A):I(B):I(C) = 1000:(145 \pm 10\%):(5 \pm 25\%)\,. \tag{29}$$

For KBr one obtains

$$I(A):I(B):I(C) = 1000:(162 \pm 20\%):(8 \pm 50\%)\,. \tag{30}$$

This is within the error of the analysis, the same as for the case of KI.

In a series of experiments it was verified that the threefold structure of the F center in-gap mode is not due to the presence of impurities or due to F aggregate centers (see Ref. [87]). This leads to the conclusion that this structure is due to the presence of two stable potassium isotopes

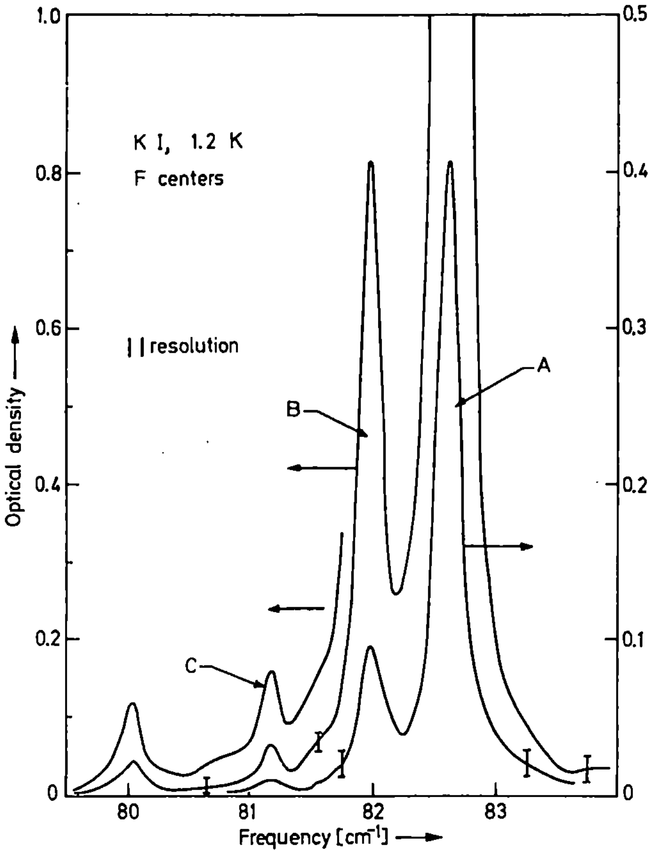

Fig. 30. F center in-gap mode in KI measured with a resolution of 0.1 cm^{-1} (apodized spectrum). The positions of the lines are: $A = 82.62 \pm 0.02$ cm^{-1}; $B = 81.98 \pm 0.02$ cm^{-1}; $C = 81.19 \pm 0.05$ cm^{-1} (after Ref. [87])

in the crystal, K^{39} and K^{41}. The natural abundance ratio of these two isotopes is K^{39} : K^{41} = 93 : 7.

One now can make the following model: The F center has three different configurations of symmetry O_h, C_{4v}, and D_{4h} (see Fig. 32). In the first case, (a), the F center is surrounded only by potassium isotopes K^{39}. Therefore, one obtains a vibrational transition from a non-degenerate A_{1g} ground state to a threefold degenerate T_{1u} excited state. The latter splits into a twofold degenerate E and a non-degenerate A_1 state when one replaces one of the 1 nn by the heavier isotope, K^{41}, thereby reducing the point symmetry to C_{4v} [case (b)]. Substituting two of the 1 nn lying on opposite sites of the F center one ends up with D_{4h} symmetry [case (c)]. The degeneracy is the same as for case (b). For case (c) the

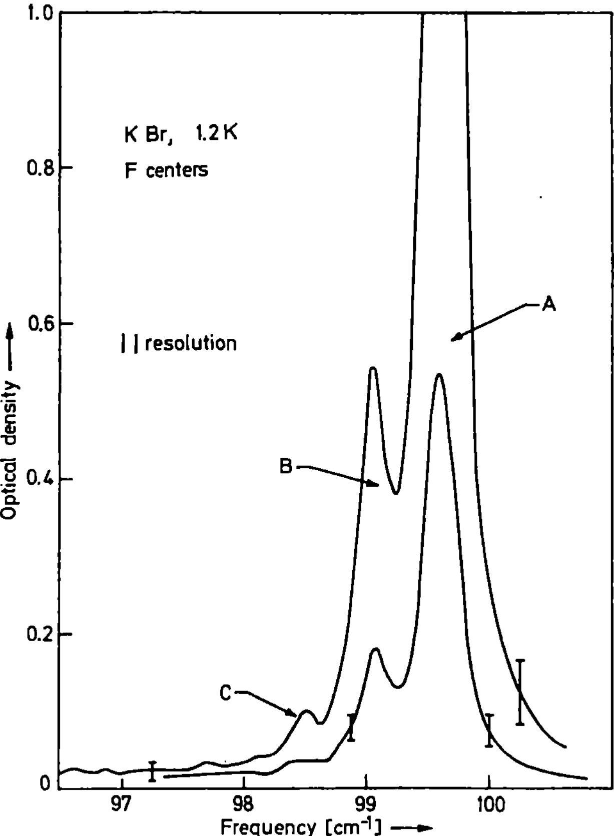

Fig. 31. Same as Fig. 30, but for KBr. The positions of lines are $A = 99.60 \pm 0.03\ \mathrm{cm}^{-1}$; $B = 99.07 \pm 0.04\ \mathrm{cm}^{-1}$; $C = 98.50 \pm 0.05\ \mathrm{cm}^{-1}$ (after Ref. [87])

splitting of the excited levels, however, is larger because of the bigger perturbation in effective masses.

One now can assume that cases (a) to (c) differ only in the mass perturbation; i.e. it is assumed that the local force constants are the same for all three cases. Then one can reduce *all* possible configurations in the crystal to cases (a) to (c). As an example, case (d) then does not lead to additional vibrational transitions. The B_1 and E, and the B_2 and A_1 vibrations respectively are degenerate. From this model one expects three vibrational transitions A, B, and C. One now can calculate their relative intensities: The total number of F centers shall be k. Denoting the probability that nF centers have $i\,\mathrm{K}^{41}$ ions as 1 nn by $P_i(n)$, the average occupation number is then

$$b(i) = \Sigma_n P_i(n) = k\binom{m}{i}(1 - 1/k)^{m-i} k^{-i}. \tag{31}$$

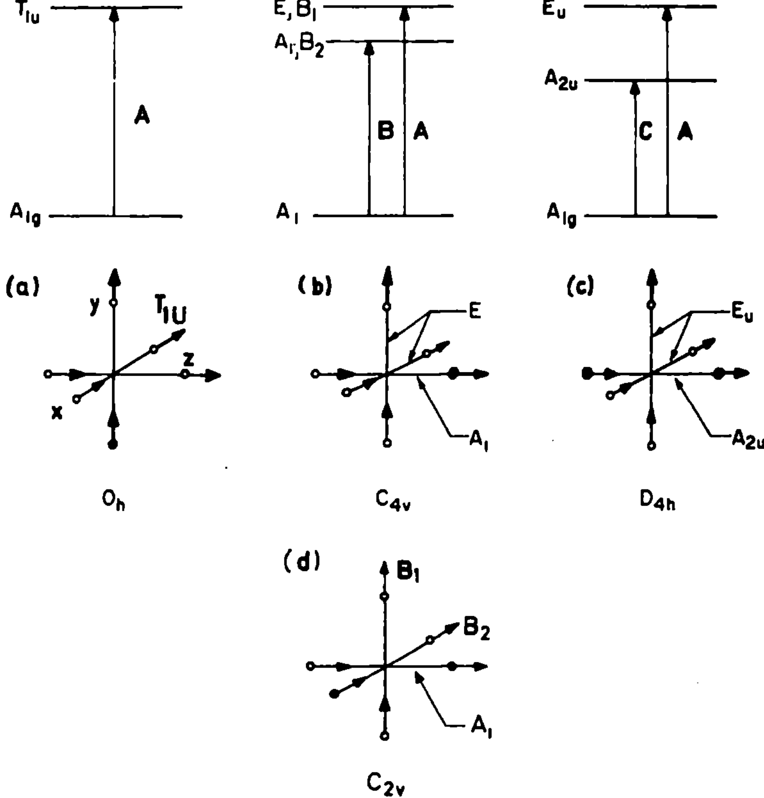

Fig. 32. Vibrational transitions for different F center configurations. Case (a): K^{39} environment only. Case (b): one of the 1 nn substituted by K^{41}. Case (c): two 1 nn on opposite lattice sites substituted by K^{41}. Case (d): two 1 nn on *non*-opposite lattice sites substituted by K^{41}

For the natural abundance ratio of $K^{39}:K^{41}=93:7$. One finds among $6k$ potassium ions $m=6k\cdot7\cdot10^{-1}$ of the heavier isotope. One then obtains

$$b(0)=65.5\%;\quad b(1)=27.7\%;\quad b(2)=5.9\%.$$

In order to compare the relative frequencies of these three configurations with the relative intensities of the lines, one must take into account the polarization and the degree of degeneracy in the modes according to Fig. 32. One then obtains

$$A=2b(0)+(4/3)\,b(1)+(12/15)\,b(2),$$

$$B=(2/3)\,b(1)+(16/15)\,b(2),$$

$$C=(12/15)\,b(2).$$

The above values then yield

$$I(A):I(B):I(C) = 1000 : 143 : 5.$$

This is in excellent agreement with the experimental data (29) and (30). This results supports the model and encourages a threedimensional model calculation.

4.2.3. Model Calculation

The calculation is based on the model of Fig. 4. Changes in the tangential force constant B are neglected. For $s = 0, 1, 4$ the symmetry vectors (see Section 2.4) are $\sigma(T_{1u}, j, 0)$; $\sigma(T_{1u}, j, 1, 1)$; $\sigma(T_{1u}, j, 4, 1)$ where $j = 1, 2, 3$.

In the following the projected perturbation matrix $\hat{V}(T_{1u})$ for configurations (a) to (c) of the F center according to Fig. 32 is discussed.

Case (a): F center in O_h symmetry

$$M' < M_A ; \qquad \varepsilon_- = 1 - M'/M_A$$

$$A01 \neq 0 ; \qquad A14 \neq 0.$$

The perturbation matrix $\hat{V}$ can be written

$$\hat{V} = \begin{bmatrix} \hat{V}_x & & \\ & \hat{V}_y & \\ & & \hat{V}_z \end{bmatrix}$$

V_α are symmetrical 3×3 matrices of the form

$$= \begin{bmatrix} \varepsilon_- \cdot \omega^2 + A \cdot A01 \cdot M_A^{-1} & -\sqrt{2}\, A \cdot A01 \cdot (M_A M_K)^{-1/2} & 0 \\ \dots\dots\dots\dots\dots & A(A01 + A14) \cdot M_K^{-1} & A \cdot A14 \cdot (M_K M_A)^{-1/2} \\ \dots\dots\dots\dots\dots & \dots\dots\dots\dots\dots & A \cdot A14 \cdot M_A^{-1} \end{bmatrix}$$

and

$$\hat{V}_x^{(a)} = \hat{V}_y^{(a)} = \hat{V}_z^{(a)} .$$

In this case all gap modes are threefold degenerate.

Using the assumption of Section 4.2.2 ($A01, A14$ the same for the three configurations) the perturbation matrix can be written directly down for cases (b) and (c).

Case (*b*): F center in C_{4v} symmetry

$$M' > M_K; \quad \varepsilon_+ = 1 - M'/M_K$$

where M' and M_K are masses of K^{41} and K^{39} isotopes respectively

$$\hat{V}_x^{(b)} = \hat{V}_y^{(b)} = \hat{V}_x^{(a)}$$

$$\hat{V}_z^{(b)} = \hat{V}_z^{(a)} + \begin{bmatrix} 0 & & \\ & \tfrac{1}{2}\varepsilon_+ \cdot \omega^2 & \\ & & 0 \end{bmatrix}$$

One gets two types of modes: the unperturbed modes of the first model (twofold degenerate) and one non-degenerate new mode.

Case (*c*): F center in D_{4h} symmetry

$$\hat{V}_x^{(c)} = V_y^{(c)} = \hat{V}_x^{(b)} = \hat{V}_x^{(a)}$$

$$\hat{V}_x^{(c)} = 2\,\hat{V}_z^{(b)} - \hat{V}_z^{(a)} = \hat{V}_z^{(b)} + \begin{bmatrix} 0 & & \\ & \varepsilon_+ \cdot \omega^2 & \\ & & 0 \end{bmatrix}$$

The degeneracy is the same as for case (b), but the splitting is larger. No further gap modes can occur in connection with the heavier mass isotope K^{41}. All T_{1u} modes which strain the springs between 1 nn and 2 nn correspond to the case of a diatomic linear chain with $M_K < M_A$ and $\varepsilon_+ < 0$, where no gap modes exist. For a more complete discussion see Refs. [87] and [20] .

4.2.4. Numerical Results for Frequency Positions

Figures 33 and 34 show the calculated frequency dependence of lines A, B, and C of the F center in-gap mode in KI and KBr as a function of force constant changes $A01$ and $A14$. The results are in agreement with the predictions in Section 2.3: the line positions depend strongly on $A14$, the force constant change between 1 nn and 4 nn. For any *fixed* value of $A01$ the peaks may be shifted over the total gap region by a variation of $A14$ within reasonable limits $(-0.3 \leq A14 \leq 0.3)$. On the other hand the "rigid oscillator model" in which the F center and 1 nn vibrate against 4 nn, is too simple. The frequency positions depend also on $A01$ – but by a much smaller amount.

The threefold structure of the F center in-gap mode allows a quantitative determination of the force constant pairs, $A01$ and $A14$, which is

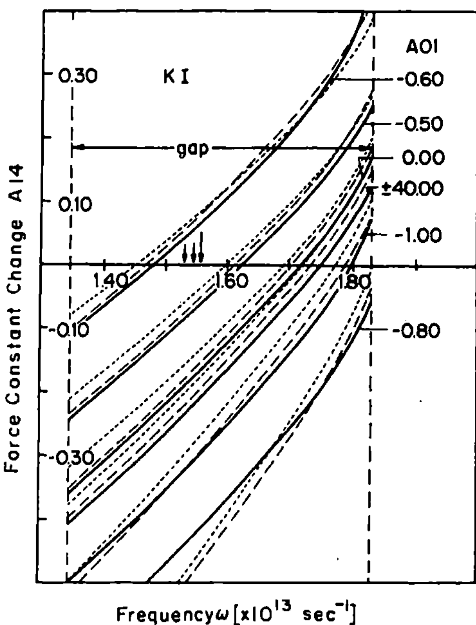

Fig. 33. Dependence of the line position and splitting of the F center in-gap mode absorption in KI as a function of force constant changes $A01$ and $A14$. Solid line: F centers in the $b(0)$ configuration. Dashed line: F centers in the $b(1)$ configuration. Dotted line: F centers in the b (2) configuration. The calculation was done for 0 K. The arrows show the experimental positions of lines (after Ref. [87])

independent of data from the electronic F band: a fit of the observed frequency for the line A in KBr and KI restrict the possible force constant pairs $A01$, and $A14$. A further restriction is obtained by fitting the measured splitting of lines A and B. The frequency of line C is best reproduced by choosing values as listed in Table 14. The positive sign for $A14$ for KBr is at the limit of accuracy of the calculation. However, one can see the tendency of $A14$ to decrease when going from KBr to KI. In contrast, $A01$ remains roughly constant!

A comparison of local force constant changes for F centers (Table 14) and U centers (Table 7) is not straight forward. The reason is that the force constant changes may "largely" depend on the model used for calculating host lattice eigenvalues and eigenvectors. Keeping this in mind one can at most speak about the tendency of local force constant changes when substituting U centers with F centers: From Tables 7 and 14 one finds that values $A01$ for U centers and F centers may be comparable (this was also found in the calculations by Singh and Mitra [95]; see also Section 4.2.1) and only "slightly" different for different substances

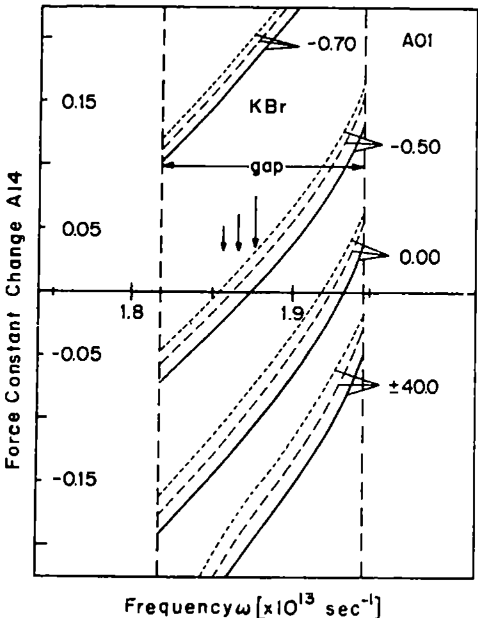

Fig. 34. Same as Fig. 33, but for KBr (after Ref. [87])

Table 14. Local force constant changes $A01$ and $A14$ for F centers in alkali halides

Substance	$A01$	$A14$
KBr	-0.50	$+0.002$
KI	-0.50	-0.060

Force constant changes obtained from a defect model calculation — based on Schröders breathing shell model-fitted to frequencies of lines A, B, and C of the F center in-gap mode.

(the case of NaF is excluded). On the contrary, values $A14$ seem to be much smaller for U centers than for F centers, and decrease with increasing size of halide ions. A possible heuristic explanation may be the following: The force constant $A14$ is determined by the relaxation of 1 nn, while $A01$ is mainly due to the "vacancy", i.e. an additional contribution due to relaxation is small. The relaxation of 1 nn ions is governed by short range Born-Mayer repulsive forces and by Coulomb forces. The repulsive forces are smaller between the defect and 1 nn than that between 1 nn and 4 nn. This leads to an inward relaxation of 1 nn ions. The Coulomb forces are determined by the charge within the first shell. For U centers the charge is completely localized within the first shell.

For F centers, however, there is a depletion of charge due to the locally large extended ground state, i.e. a positive "defect charge" within the first shell occurs. This results in a Coulomb repulsion, i.e. an outward relaxation of 1 nn, and thereby in an increase of $A14$. This would explain the smaller values $A14$ for U centers relative to the values for F centers. In addition the larger value of $A14$ for F centers in KBr as compared to that for KI could be explained in this manner. The depletion of charge is larger with KBr than with KI [the lattice constants are a (KBr) = 3.28Å, a(KI) = 3.48Å]. For KCl with F centers one would expect the value of $A14$ to be even larger than for KBr. This is supported by an analysis of ENDOR measurements by Kersten [99], which yields the relaxation of 1 nn. From the simple model suggested in Ref. [87], one calculates with Kersten's data $A14 = +0.072$ for KCl.

A further support of these arguments is obtained from experiments with charged defects like F' centers (see Section 4.5).

4.2.5. Stress Experiments

The stress splitting of lines A and B of the F center in-gap mode in KI is shown in Fig. 35 for $E \parallel P \parallel [100]$ and $E \parallel [001]$, $P \parallel [100]$. E is the electric vector of the incident light, and P the stress. Each line splits into two components which shift linearly with applied stress but by different amounts. Within the accuracy of measurements no change in the band shape or the integral absorption ($\pm 30\%$) was observed. From the slope of the function $\Delta v_{1,2}(A) = f(P)$, one can, for different directions of polarization, determine the stress coefficients α and β (see Section 3.1.5).

Uniaxial stress in the [100] direction lowers the point symmetry for F centers in the $b(0)$ configuration from O_h to D_{4h}. In the $b(1)$ configuration the symmetry becomes C_{4v} for the c axis $\parallel P$ and C_{2v} for the c axis $\perp P$ (see Fig. 36). Centers in the b (2) configuration shall be neglected in the following. However, because no measurable splitting was observed between the A_{2u} and B_1 levels and among the E_u and E and B_2 levels, one can assume that the static distortions are nearly of A_{1g}, E_g, and T_{2g} symmetry.

The local compliances were calculated according to Benedek and Nardelli [85]. With the force constant changes $A01$ and $A14$ determined in 4.2.4 one obtains

$$s_{11} + 2s_{12} = 3.05 \times 10^{-12} \text{ cm}^2/\text{dyn},$$

$$s_{11} - s_{12} = 5.00 \times 10^{-12} \text{ cm}^2/\text{dyn}.$$

From Eq. (25) one calculates for line A:

$$\alpha(A_{1g}) = 125 \pm 40 \, cm^{-1}; \quad \beta(E_g) = 35 \pm 15 \, cm^{-1}. \tag{32}$$

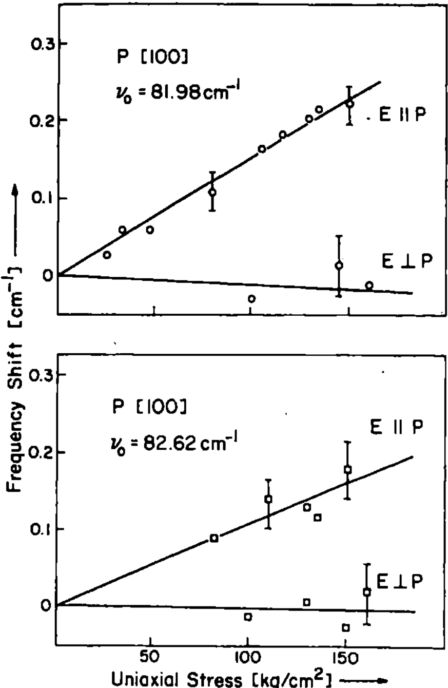

Fig. 35. Frequency shift $\Delta v = v - v_0$ of the lines $A(v_0 = 82.62\ \mathrm{cm}^{-1})$ and $B(v_0 = 81.98\ \mathrm{cm}^{-1})$ of the F center in-gap mode in KI (5 K) with uniaxial stress (after Ref. [87])

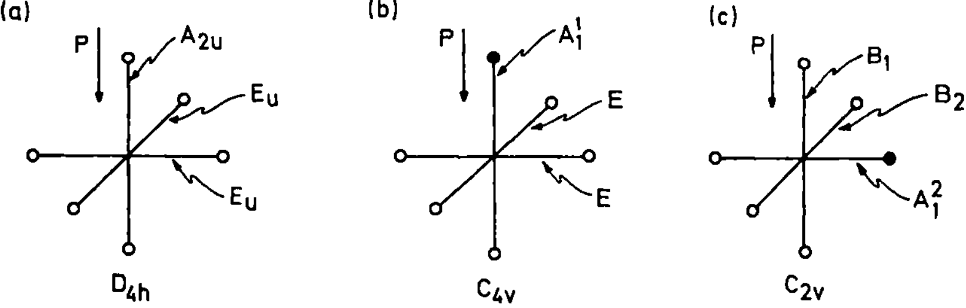

Fig. 36. Behaviour of F centers in the b (0) [case (a)] and b (1) [cases (b) and (c)] configurations with uniaxial stress. O denotes K^{39} and ● denotes K^{41} isotopes. The single components of the split lines A and B contain

$$P \parallel E : v_1(A) : A_{2u};\ B_1$$
$$v_1(B) : A_1^1\ \text{"shifted"}.$$

$$P \perp E : v_2(A) : E_u;\ E;\ B_2$$
$$v_2(B) : A_1^2\ \text{"slightly shifted" (after Ref. [87])}$$

This result is consistent with the experimental values for line B. The excited vibrational state of the F center is coupled about three times more strongly to spherical distortions A_{1g} then to tetragonal distortions E_g. This ratio is similar to that for U centers.

Because the half widths of lines A, B, and C of the F center in-gap mode were not yet resolved, an estimate of the third order anharmonic coefficient – which could be compared with the stress coefficients – is not possible.

4.3. $F_A(\text{Na})$ Centers

Preliminary experiments have been performed on the vibrational absorption of $F_A(\text{Na})$ centers [86] (see Section 2.1). Besides the unperturbed F center in-gap mode two additional lines in the difference spectra were found at $80.05 \pm 0.04\,\text{cm}^{-1}$ and at $72.61 \pm 0.04\,\text{cm}^{-1}$ [4]. In principle these lines could be correlated with vibrational transitions $A - A_1$ and $A_1 - E$ [see Fig. 32 case (b) – in analogy with electronic transitions F_{A1} and F_{A2}]. The intensity ratio of these additional lines seems to be in rough agreement with the intensity ratio expected from Eq. (31) when taking into account the polarization and the degree of degeneracy in the modes. However, a further check of the correlation of these lines with $F_A(\text{Na})$ centers is necessary. Most illuminating would be experiments on the dichroism at the F_A center far infrared lines.

4.4. M Centers

In additively coloured crystals F centers were converted to M centers (at $300\,\text{K}$ by irradiation of the crystal with F center light). No sharp absorption line of similar oscillator strength, when compared with the F center mode, was observed in the gap.

4.5. Infrared Vibrational Absorption of F' and α Centers

In the preceeding sections defects whose valency is the same as that of the ion replaced have been discussed.

In this chapter the influence of the effective charge within the "vacancy" on the infrared vibrational frequency $T_{1u}(2)$ will be studied – at least experimentally. This can be done by converting F centers to F' centers and α centers, or U centers to U_1 centers and α centers (see Fig. 1, and Section 2.1).

---

[4] In Na doped KI, in addition to the defect induced acoustic band mode absorption earlier seen by Sievers [18], an in-gap mode was found at $83.81\,\text{cm}^{-1}$, which is due to a localized mode of Na^+ centers.

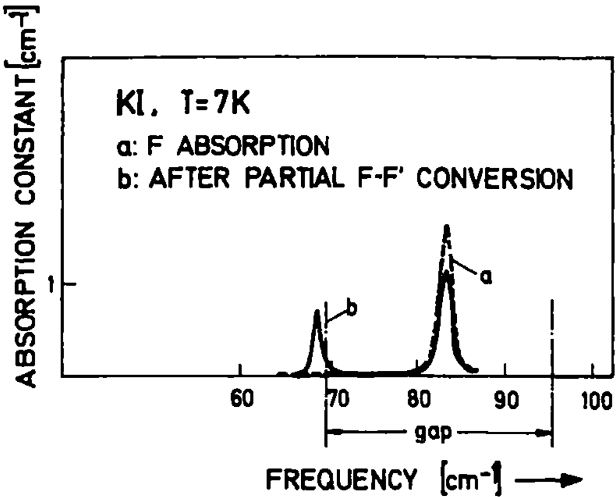

Fig. 37. $F - F'$ conversion in KI at 7 K (after Ref. [91])

The result of the $F-F'$ reaction in KI is shown in Fig. 37. A fraction of about 30 % of F centers was converted to F' centers and anion vacancies, α centers. In the infrared spectra a new resonant mode at the upper edge of the acoustic band near 68.5 cm^{-1} occurs. This line is due to F' centers[5] This could be verified by using the $U-\alpha$ process (see Section 2.1). This reaction leads to a new absorption line near 87 cm^{-1}, which can be correlated with interstitial hydrogen ions, $U_{\rm I}$ centers (see Section 4.6). No absorption occurs at 68.5 cm^{-1}. α centers do not absorb in the frequency region investigated.

Qualitatively, one can understand these spectra from the results obtained in Section 4.2. An additional charge within the vacancy attracts 1 nn while 4 nn are repelled [see Fig. 38 (a)]. This results in a decrease in the force constant A 14. According to Fig. 33 one therefore expects a lower frequency for the vibrational absorption of F' centers when compared with the frequency of the F center in-gap mode.

The anion vacancy behaves as a defect charge. Therefore, one has the opposite conditions [see Fig. 38 (b)]. 1 nn are repelled while 4 nn are attracted. This results in an increase in A 14. An absorption due to α centers is therefore expected at a higher frequency relative to the F center in-gap mode. Because no absorption due to α centers was found one can assume that this absorption is covered by the intrinsic absorption of the host lattice.

For a quantitative analysis of the line position of the infrared vibrational absorption due to F' centers, one can no longer assume trans-

---

[5] The instrumental resolution in this experiment was about 1 cm^{-1}. A high resolution measurement on this line — in analogy to the measurements on the F center in-gap mode (see Section 4.2.2) — would be of great interest for a quantitative treatment of the problem.

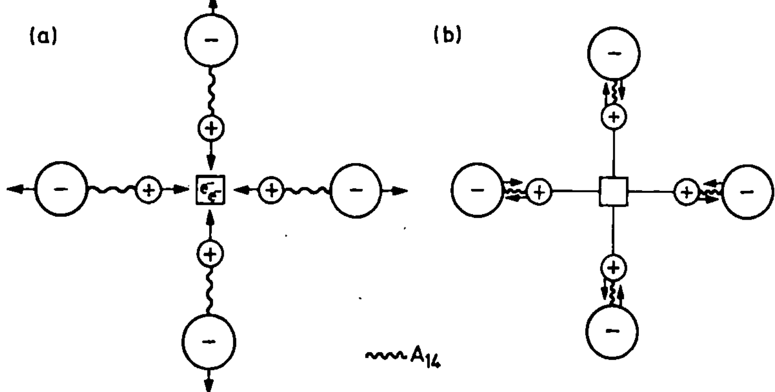

Fig. 38. Model for relaxation of 1 nn of the F' center [case (a)] and the α center [case (b)]. Arrows indicate directions of relaxation

lational symmetry in the matrix of effective charges. Furthermore, it seems to be doubtful whether a dynamical model which is limited to a small region around the defect (see Section 2.4) is appropriate at all. On the contrary, one would assume that the additional charge within the vacancy causes Coulomb potential changes which extend far into the lattice. Under these aspects a theoretical treatment of the F' center vibrational problem seems to be exciting.

4.6. U_1 Centers (H_i^-, D_i^- Centers)

In the far infrared spectra of KBr and KI the $U-\alpha$ process yields additional absorption lines at $98.7\pm0.5\,\mathrm{cm}^{-1}$ and at $86.7\pm0.5\,\mathrm{cm}^{-1}$, respectively (see Figs. 23 and 39). Because F centers are simultaneously produced (see Section 2.1), the F center in-gap mode is also present in both spectra. The correlation of the additional absorption lines with U_1 centers was established by using the $F-F'$ reaction (see 4.5). Furthermore, the lines bleach along with the ultraviolet U_1 band, which was controlled in the same experiment. The thermal stability of these centers corresponds to "free" U_1 centers (see Section 3.2). The half widths of the lines are $\lesssim 1.8\pm0.5\,\mathrm{cm}^{-1}$ and $\lesssim 1.3\pm0.5\,\mathrm{cm}^{-1}$ for KBr and KI, respectively. Substituting H_i^- centers for D_i^- centers the same lines were obtained. Within the accuracy of the measurement no frequency shift was found.

The nature of the U_1 center in-gap mode is quite different from the F center in-gap mode or the U center resonant mode. No simple explanation on the basis of an altered mass and altered force constants

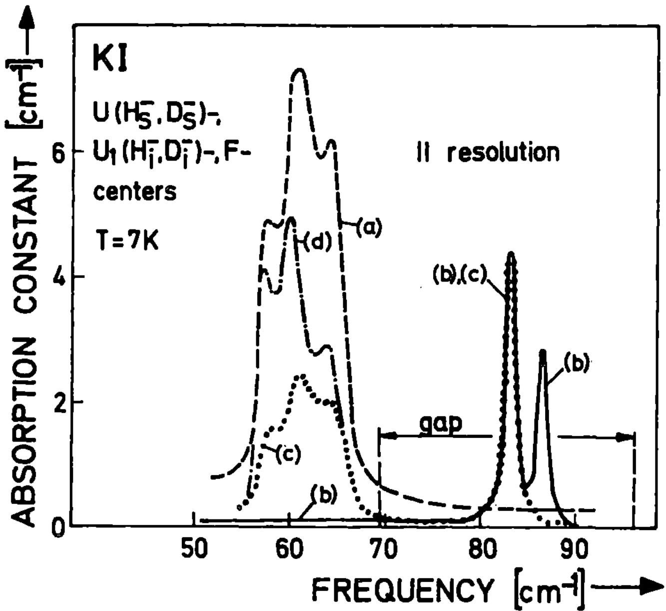

Fig. 39. Far infrared spectra at 7 K. (a) H_s^- centers (4×10^{18} cm^{-3}). (d) D_s^- centers. (b) after mercury arc irradiation (U_1, α, and F centers present), (c) after thermal recombination of α and U_1 centers (after Ref. [71])

can be given, since three new degrees of freedom are added to the section of the lattice which contains the defect. It is known, however, that the interstitial ion introduces a distortion of its cage of nearest neighbors [61]. It is proposed that this favors collective motions of the whole body of the interstitial ion and its four anion neighbors; a resulting resonant or localized mode displacement is shown in Fig. 41.

In Section 3.2 it was mentioned that infrared active modes of T_2 symmetry can couple to the localized mode. In Figs. 40 and 41 near infrared and far infrared spectra are compared. It is evident that the frequency of the strongest line in the sideband spectrum of the high frequency localized mode agrees quite well with the position of the line in the far infrared. This leads to the conclusion that the strong sideband peak is caused by a phonon, which also can be directly excited in the far infrared; i.e. one can make the correlation (see Section 3.2)

$$\Gamma = T_2^P = T_2(2)$$

This means that one can analyze the sideband spectrum of the high frequency localized mode with respect to T_2 peaks by investigating the far infrared spectra. Thereby, direct information on the coupling co-

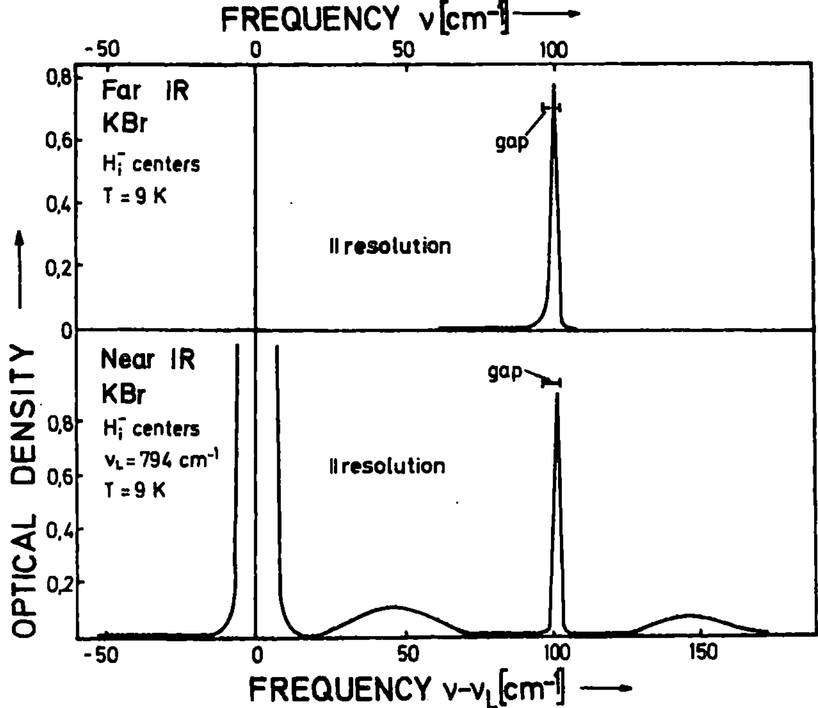

Fig. 40. Far infrared and near infrared vibrational absorption due to H_i^- centers in KBr (after Ref. [72])

efficient $A(\Gamma)$ is obtained. Using Eq. (28) and Eq. (23), employed for T_d symmetry, one finds

$$A(T_2(2)) \approx K \sum_{\Gamma} A(\Gamma)$$

where $K = 0.2$ and 0.5 for KI and KBr, respectively. If the integral absorption of the second line in the doublet in the KI spectrum is included, one would obtain $K \approx 0.4$. This would mean that the threefold degenerate $T_2(2)$ mode splits when it couples to the $T_2(1)$ localized mode. Whether this is the case or not could be proved e.g. by uniaxial stress experiments. The result that the coupling of the $T_2(2)$ mode is much stronger than the coupling of other modes is quite different from the case of the high frequency localized mode due to U centers for which coupling to A_{1g} modes is about three times stronger than coupling to modes of E_g and T_{2g} symmetry.

Another interesting effect is the disagreement of the precise far infrared and sideband peak positions ($102\,\mathrm{cm}^{-1}$, and $98.7\,\mathrm{cm}^{-1}$ for

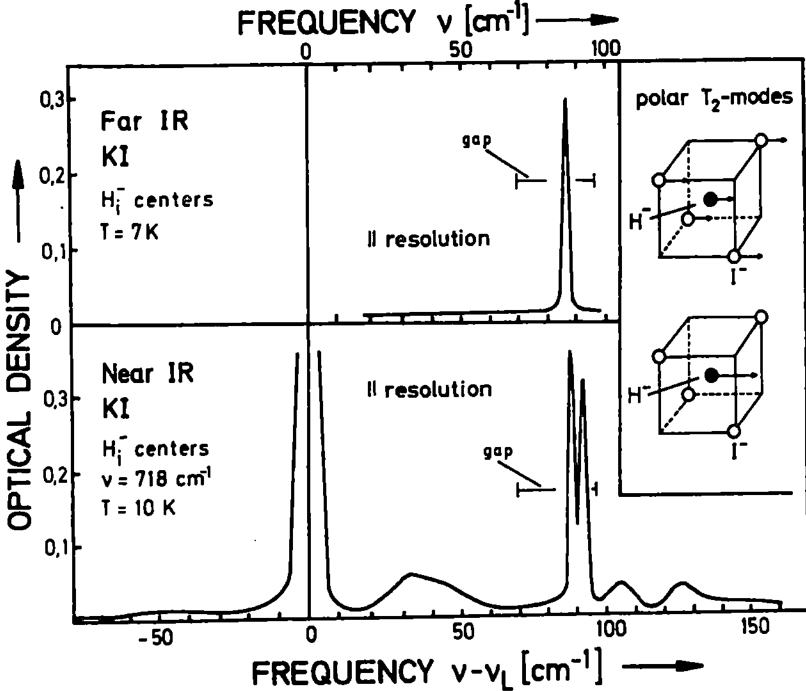

Fig. 41. Same as Fig. 40 but for KI. Insert shows model for vibrations $T_2(2)$, and $T_2(1)$ respectively

KBr, respectively). This disagreement is due to the frequency dependence of the anharmonic shift $\Delta_L(\omega)$ (see Section 3.1.2). This shifts the sideband by an amount which corresponds neither to the "free" nor to the "damped" far-infrared phonon frequency. This effect has been quantitatively treated for the U center high frequency localized mode [59].

Model calculations on the infrared vibrational absorption of U_1 centers in alkali halides were not yet performed. They could be done in analogy to calculations on the infrared vibrational absorption due to substitutional point defects by using the molecular Green function method.

5. Raman Spectra

All applications of Raman scattering to crystal lattices with point defects have taken place within the last 10 years. The reason is that high power lasers are necessary and the concentrations of the defect usually under investigation are low. In cubic crystals like alkali halides with point defects, Raman scattering is an especially powerful tool. The reason is that first order scattering is forbidden in the perfect lattice (see Section 2.2), i.e. any observed first order spectra are due only to the addition of the defect. When the defect is at a site of inversion symmetry, the only lattice modes coupled to the (virtual) electronic transition are of even parity. This means that in cubic crystals Raman scattering experiments yield information on that part of the phonon spectrum which cannot be directly excited in the infrared. Therefore, the Raman spectra complement the infrared absorption spectra.

Up to now, most Raman scattering experiments were performed in such a way that the energy of the exciting light was far from the electronic dipole transition (off-resonant or non-resonant Raman scattering). In this case one has to deal with a virtual electronic transition corresponding to a polarization of the electron shell. Since the electronic polarization consists of a mixing of the ground state with the optical-active excited states, the vibrational modes and the coupling coefficients involved are the same as those assisting the absorption process (see Section 2.2). The polarizability is assumed to be a constant, and independent of the frequency of the exciting light.

Special features occur when the energy of the exciting light, coincides with the electronic transition (resonant Raman scattering) [101, 102]. In this case the electronic polarizability can no longer be treated as a constant but may become very large and dependent on the frequency of the exciting light. The most characteristic feature of resonant Raman scattering for point defects in cubic crystals is the appearance of multiphonon spectra.

Section 5.1. reviews the experimental results on the non-resonant Raman scattering from U centers in the alkali halides KBr and KI [76], and in alkaline earth fluorides [66, 77]. Theoretical investigations were performed for U centers in alkali halides [24, 25]. Raman scattering from substitutional hydrogen centers in LiD was recently reported [103].

Near-resonant and resonant Raman scattering was observed for F centers in various alkali halides [102, 104, 105]. The present theoretical studies are based on off-resonance selection rules [27, 94, 105, 106]. A survey on these investigations is given in Section 5.2. Experimental results for F_A(Li) centers in KCl are mentioned in Section 5.3.

5.1. U Centers (H_s^-, D_s^- Centers)

In Section 3.1.4 it was found from symmetry arguments that the only phonon modes which can couple to the high frequency localized mode are of A_{1g}, E_g, and T_{2g} symmetry. These modes can couple to the electronic transition of the U center as well and should be observable in Raman scattering. In the most recent experiments by Montgomery et al. [76], however, only a weak scattering peak of A_{1g} symmetry was observed in KI with U centers. This line was found in the acoustic optical phonon gap in agreement with earlier observations in the sideband spectra [37]. In addition, well resolved second harmonics of the infrared active localized mode were found. The results for KBr: H_s^- and KI: H_s^- are shown in Figs. 42 and 43. In each spectrum three lines occur. These lines correspond to transitions from the vibrational ground state $(n=0)$ to the threefold split sixfold degenerate second excited state $(n=2)$ [see Fig. 6(a)].

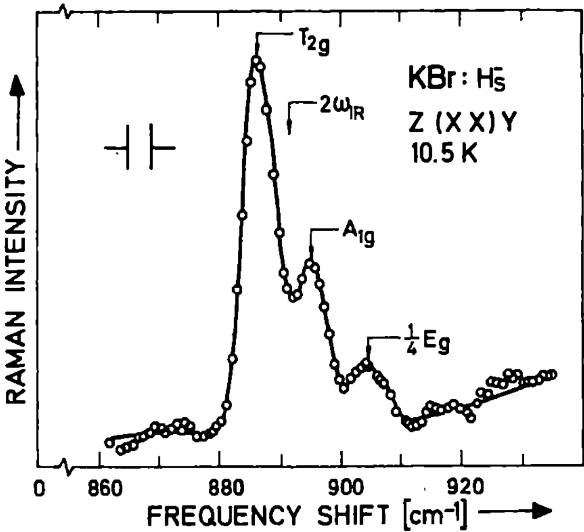

Fig. 42. $z(xx)y$ Raman spectrum of KBr: H_s^- in second harmonic region. $x \parallel [110]$, $y \parallel [\bar{1}10]$, and $z \parallel [001]$ (after Ref. [76])

The measured Raman frequencies may serve to calculate the coefficients C_1, C_2, and Ω of Eq. (15) ($B=0$ for O_h symmetry). Their values are included in Table 5. They allow one to calculate the complete energy level scheme for the U center oscillator [see Fig. 6(a)]. As examples, Table 15 shows the calculated fundamental transition frequencies $(n=0$ to $n=1)$ for KBr and KI with U centers. The good agreement with the experimentally determined infrared line positions (listed in parantheses) again indicates that the static well approximation, on which Eq. (15) is based, is a satisfactory approach for the U center localized oscillator, and yields a consistent description of infrared and Raman data.

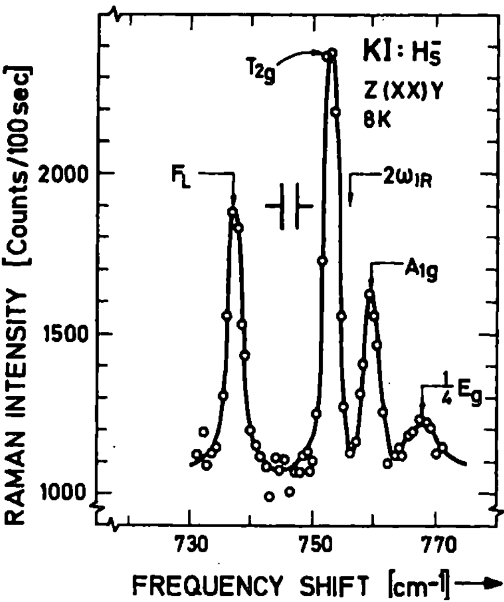

Fig. 43. $z(xx)y$ Raman spectrum of KI : H_s^- in second harmonic region. $x \parallel [110]$, $y \parallel [\bar{1}10]$, and $z \parallel [001]$. F_L denotes a plasma line of the laser (after Ref. [76])

Table 15. Calculated main line frequency for U center localized mode in KBr and KI (after Ref. [76]). Values determined by infrared absorption are in parantheses

Substance	Frequency $\nu_L(H_s^-)$ [cm^{-1}]	$\nu_L(D_s^-)$ [cm^{-1}]	Temperature [K]
KBr	446.5 (446)	319.1 (319)	11 (90)
KI	379 (382)	273 (280)	8 (90)

Table 16 lists the absolute Raman-scattering efficiencies and cross sections, which have been determined for the impurity induced modes in KI:H_s^-. It appears that the first order Raman scattering for the A_{1g} in-gap mode and the two-quantum scattering for the three harmonics is of comparable strength. As a possible explanation Montgomery et al. mention that the H_s^- ion might be a too weak electronic perturbation to produce a large first-order polarizability change. The relative intensities of the second harmonic peaks are not yet understood. They are in

Table 16. Absolute Raman efficiencies and differential cross sections for H_s^- centers in KI for 4880-Å exciting light (after Ref. [76])

| | Second harmonics | | | In-gap mode |
	T_{2g}	A_{1g}	E_g	(A_{1g})
Raman efficiency $[\times 10^{-12}\,\mathrm{cm}^{-1}\,\mathrm{sr}^{-1}]$	2.23	0.906	1.33	1.11
Differential cross Section $[\times 10^{-30}\,\mathrm{cm}^2/\mathrm{sr}^{-1}]$	3.33	1.35	1.99	1.66

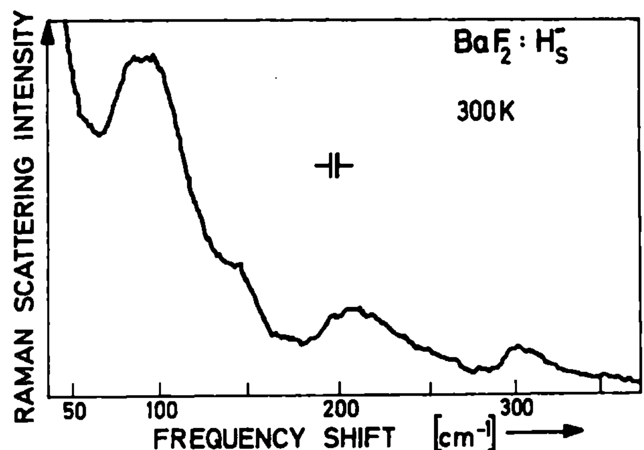

Fig. 44. High frequency sideband structure of the localized mode fundamental in $BaF_2 : H_s^-$ $(9.1 \cdot 10^{19}\,\mathrm{cm}^{-3})$ at 300 K as observed by Raman scattering. The scattering geometry displays T_2 components of the Raman tensor only. The spectrum was recorded with a time constant of 200 sec (after Ref. [66])

contradiction to the calculations which predict a much stronger scattering from $A_{1g} + E_g$ modes than from the T_{2g} mode.

An accurate Raman spectrum of localized mode sidebands was obtained in $BaF_2 : H_s^-$ by Harrington et al. [66]. Because in alkaline earth fluorides the site symmetry of U centers is T_d (see Section 3.1.1), the Raman tensor has components of A_1, E, and T_2 symmetry. The 300 K T_2 Raman spectrum of the sidebands is shown in Fig. 44; it is almost identical to the infrared spectrum of Fig. 12. This result is not surprising, since the dipole operator also transforms according to T_2 in T_d symmetry (see Section 2.2). Sideband scattering of A_1 and E symmetry was too weak to be observed in these experiments.

5.2. *F* Centers

Resonance enhancement, which occurs when irradiating near the electronic F band, allows one to obtain large Raman cross sections, even with relatively low concentrations of F centers (some 10^{17} to 10^{18} F centers cm^{-3}). This has been demonstrated experimentally by Worlock and Porto [104], Buchenauer et al. [105] and by Fitchen and Buchenauer [102] in a series of crystals like NaCl, NaBr, KF, KCl, and RbF with F centers. The F center induced Raman spectra are expected to be due to those lattice vibrations which interact with the excited electronic state. Assuming that the electron-phonon interaction is linear, and that the scattering center has octahedral symmetry, Henry and Slichter [108] have shown that the near resonance first-order Raman spectrum contains only modes of A_{1g}, E_g, and T_{2g} symmetry (see Section 2.2). These are modes which determine the width of the electronic F band.

The F center induced Raman spectra for NaCl, measured by Worlock and Porto, are shown in Fig. 45. The figure includes theoretical results for the first-order Raman spectra obtained by Benedek and Mulazzi [94] for the model of Fig. 4. For A_{01} and A_{14} the fitted values were taken (see Section 4.2). The i_{ij} are components of the first order Raman tensor given in Section 2.4. Because for NaCl $\beta \ll \alpha$ one has $i_{11} \approx i_{12}$.

From the comparison of experimental and theoretical data some interesting statements can be made: Firstly, discrepancies $\Delta\Omega$, between main peaks of observed and calculated curves occur. These were interpreted as being due to anharmonic contributions. They arise mainly from strong asymmetry in the potential of 1 nn of the F center. Secondly, the experimental data continue beyond the one phonon density of states, i.e. multiphonon scattering can not be neglected.

Even stronger multiphonon spectra occur for resonant Raman scattering. Experimentally this was shown by Fitchen and Buchenauer [102]. Theoretically Rebane et al. [109] found that the total m-phonon Raman cross sections follow the relation

$$(d\sigma/d\Omega)_m \propto (\hbar\omega_i)^4 |\phi_m(z)|^2 \tag{33}$$

where

$$\phi_m(z) = (2/m!\,2^m)^{1/2}\,(d/dz)^m \left[(\pi/2)^{1/2} + i\int_0^z e^{x^2}\,dx\right] e^{-z^2}$$

and

$$z = \hbar(\omega_i - \omega_F)/\sqrt{2}\,\sigma\,.$$

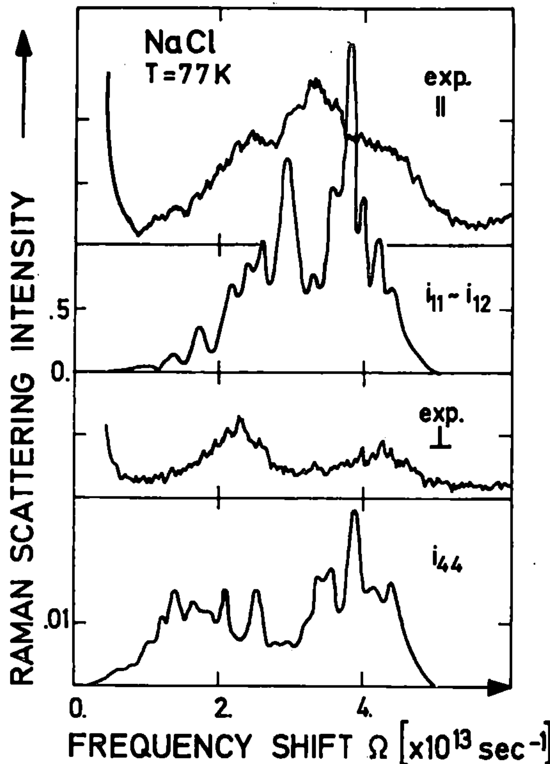

Fig. 45. Raman scattering intensity versus frequency shift for the F center in NaCl. The experimental data and the theoretical results, which were obtained by using fitted force constants for A_{01} and A_{14}, are compared (after Ref. [94])

σ is the second central moment of the band. z is the separation of the exciting (laser) frequency, ω_i, from the absorption band frequency, ω_F (in our case the center frequency of the electronic F band), in terms of the halfwidth of the band. The variation of $\phi_m(z)$ with z is shown in Fig. 46. This figure shows two remarkable features: first, the enhancement of scattered intensity as one approaches resonance, and second, the fact that multiphonon scattering intensities fall off slowly at resonance, but rapidly far from resonance. For comparison the assumed Gaussian shape of the absorption band is shown by the dashed curve.

The role of multiphonon scattering can be seen for the case of NaBr with F centers from Fig. 47. The most remarkable feature is a strong resonance peak at $136\,\mathrm{cm}^{-1}$ which occurs just above the phonon gap which extends from $105\,\mathrm{cm}^{-1}$ to $127\,\mathrm{cm}^{-1}$. Harmonics of this peak can be seen up to fourth order, and are indicated by the arrows. They can be

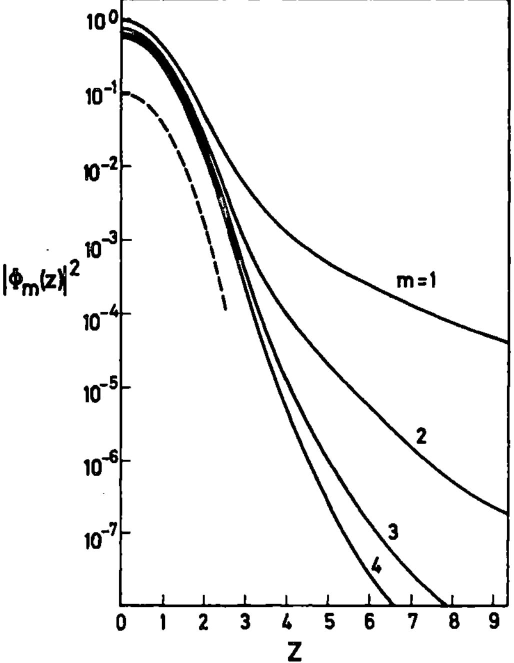

Fig. 46. Variation of the mth-order scattered intensity with separation of the incident frequency from the peak absorption frequency (after Ref. [102])

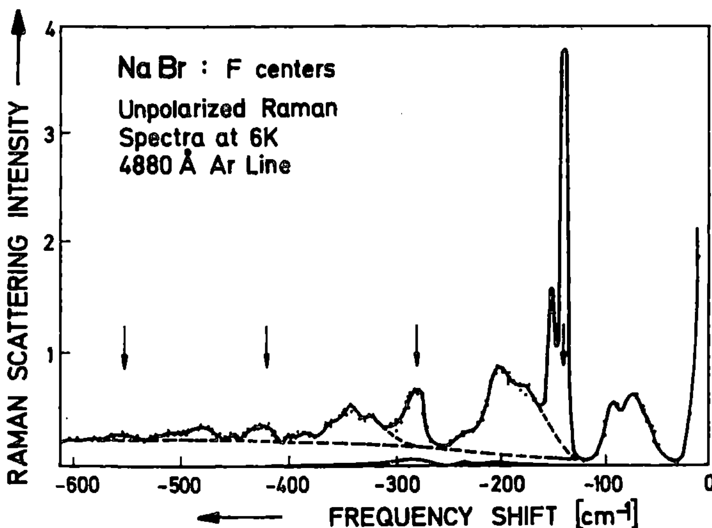

Fig. 47. Unpolarizes Stokes Raman spectrum at 6 K for F centers in NaBr with 4880Å excitation. The dashed lines suggest separation into different orders. The solid curve at the bottom is 2nd order scattering by pure crystal (after Ref. [102])

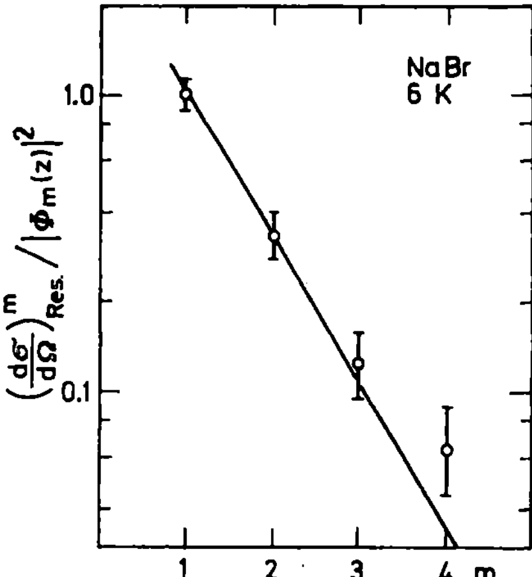

Fig. 48. Predicted and measured behavior of $(\sigma_i/\sigma_{total})^m$ for the $136\,cm^{-1}$ A_{1g} peak in NaBr with F centers (after Ref. [102])

used for testing Eq. (33) if it is generalized to the case of several coupled modes. This involves an additional factor $(\sigma_\Gamma/\sigma_{Total})^m$, which represents the relative scattering cross section for mode Γ. This factor can be determined independently from the integrated intensity of the peak and the total intensity in first order in different polarizations. Figure 48 shows the observed intensities up to fourth order. The straight line indicates the experimentally determined factor $(0.32)^m$.

A theoretical analysis of the NaBr Raman data was performed by Benedek and Mulazzi [106]. Off resonance selection rules were employed. The extended model was used and second order processes were taken into account. Figure 49 shows the results. The projected phonon density $\varrho(A_{1g})$ and its convolution $\varrho(A_{1g}) \times \varrho(A_{1g})$ give the main contributions to the first and second order parallel spectra, respectively. The peak in the $\varrho(A_{1g}) \times \varrho(A_{1g})$ term is just the overtone of the $136\,cm^{-1}$ line. For $\perp$ spectra the terms $\varrho(T_{2g})$ and $\varrho(A_{1g}) \times \varrho(T_{2g})$ are shown. Both, parallel and perpendicular Raman spectra are reproduced quite well. The superposition of the projected one and two-phonon densities was chosen in a way to fit the experimental data as good as possible. The coefficients of coupling for the electronic transition to lattice vibrations can . be estimated from weighting factors. Unfortunately experimental stress data for NaBr with F centers are not yet available for comparison. The fitted force constants are $A01 = -0.67$, and $A14 = -0.25$.

The occurrence of the resonant mode in NaBr with F centers is in analogy to the results in Section 4.2.1 and 4.2.4. Its origin is due to large

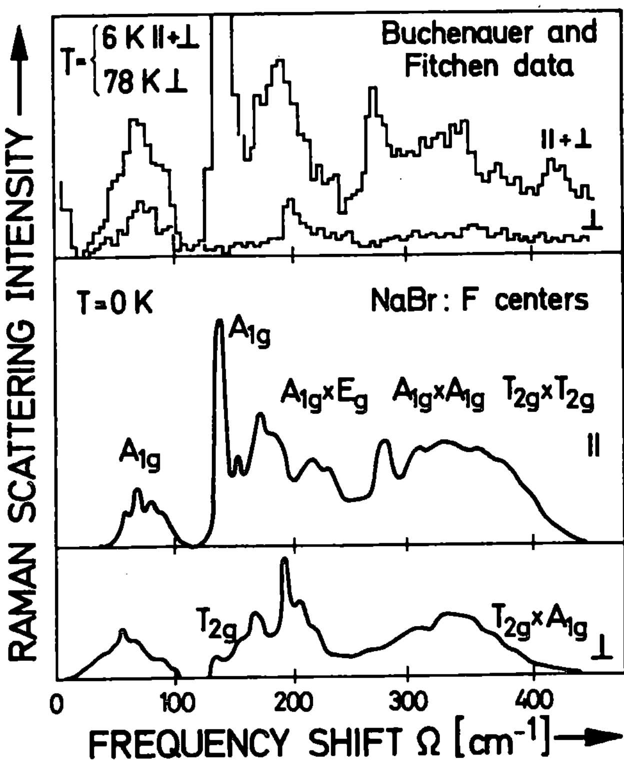

Fig. 49. Raman scattering intensity versus frequency shift for the F center in NaBr. The experimental data and the theoretical results, which were obtained by using fitted force constants for $A01$ and $A14$, are compared (after Ref. [106])

changes in local force constants. In fact, a strong gap mode of E_g symmetry in the Raman spectra of KI with F centers is predicted as well [94]. This is not yet measured.

New features occur in the resonant Raman spectra of KCl with F centers (see Fig. 50). A considerable change in scattered intensity is found when the wavelength of the excitation is varied from the F band to the K band. A broad band at about $190\,\mathrm{cm}^{-1}$ dominates the spectra for K band excitation. It is suggested that the $190\,\mathrm{cm}^{-1}$ band arises from long wavelength LO phonons. Since it is known that the excited states of the F center extend much further than the 1 nn shell, a coupling of LO phonons through the Fröhlich interaction has long been expected. K band states have radii of tens of Angströms and more. The total Raman cross section for one phonon scattering at 5145 Å is about $3 \cdot 10^{-22}\,\mathrm{cm^2/sr}$ (compare Table 16)!

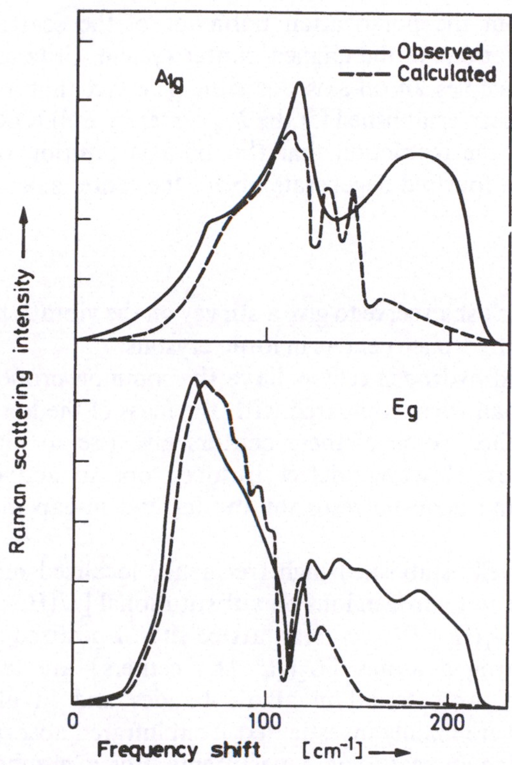

Fig. 50. Comparison of projected 1 nn phonon densities of states for perturbed modes in KCl ($A01 = -0.60$) with measured first order spectral densities for F centers in KCl at 6 K (after Ref. [110])

The calculated curves in Fig. 50 are obtained with a dynamical model based on the breathing shell model. Changes in force constant A_{01} were taken into account only. The best fit of the experimental curve outside the 190 cm^{-1} band was obtained for $A01 = -0.60$.

The relative strength of A_{1g}, E_g, and T_{2g} phonons was found to be 0.36 ± 0.05, 0.25 ± 0.04, and 0.40 ± 0.06. These values are in reasonable agreement with the relative contributions of these modes to the half width of the electronic F band as determined from stress induced dichroism.

5.3. F_A(Li) Centers

The resonant Raman scattering from F_A(Li) centers in KCl was studied by Fritz [107]. Besides changes of the spectrum in the F center scattering region, two new peaks at 288 cm^{-1} and 228 cm^{-1} (8 K) were observed. On substituting Li^7 instead of Li^6 the high-frequency line is shifted to

$268\,\mathrm{cm}^{-1}$. From the polarization behavior of the scattered light for different orientations of the aligned center system, Fritz suggested that the Li^+ ion occupies an off-axis position. The fact that an overall C_{4v} symmetry has been established for the F_A center by ENDOR experiments [111] leads to the restriction that the off-axis position of the Li^+ion must be at least fourfold degenerate about the center axis.

6. Summary

The present work shall serve to give a survey on the vibrational properties of electron and hydrogen centers in ionic crystals.

Electron and hydrogen centers have the common property that their mass is very small when compared with the mass of the host lattice ions. As a consequence, some of these centers give rise to high frequency localized modes. However, defect induced optical activity of lattice band modes, and acoustic resonant modes and in-gap modes are observed as well.

The most well established high frequency localized modes in ionic crystals are due to hydrogen ions on substitutional $[U(H_s^-, D_s^-)$ centers$]$ or interstitial $[U_1(H_i^-, D_i^-)$ centers$]$ lattice sites. Localized modes due to interstitial hydrogen atoms $[U_2(H_i^0, D_i^0)$ centers$]$ are fairly well established. They were found in alkali halides and in alkaline earth fluorides, and were mainly investigated from infrared absorption spectra and/or from Raman scattering experiments. For a number of crystals localized modes due to hydrogen ions and/or hydrogen atoms were established indirectly from vibronic spectra. They are not discussed in this work.

In alkali halides and alkaline earth fluorides the near infrared absorption spectra of U centers, and U_1 centers consist of a strong main line (fundamental) and sidebands. The main line is due to a vibrational transition from a non-degenerate ground state to a threefold degenerate first excited state. Higher harmonic vibrational transitions were observed for the case of U centers in the infrared absorption and/or Raman spectra. Their energy and splitting can be explained by an anharmonic oscillator, which is vibrating in a static potential well. The anharmonic coefficients, and thereby the complete oscillator level scheme can be calculated from the observed transition frequencies. The temperature dependent halfwidth and frequency shift of the fundamental transition, and the sideband structure observed, can be explained by anharmonic coupling of the localized mode to lattice band modes. Microscopic interaction mechanisms are suggested, and the amount of the coupling coefficients can be estimated. Experiments with uniaxial stress, electric fields, and mixed crystals support and complement the results.

In the far infrared spectra acoustic resonant modes and in-gap modes were found for U centers, F' centers, and for F centers, $F_A(\text{Na})$ centers, U_1 centers, respectively. Resonant modes eventually present within the optical band can not be directly observed in ionic crystals because of the strong "reststrahl-absorption".

Raman scattering experiments from F centers in alkali halides did yield information on the density of even parity lattice modes and on their relative amount of coupling to the electronic transition.

From the number of vibrational transitions observed in the infrared and/or Raman spectra the local site symmetry of a defect can be determined. Its local force constants and vibrational amplitudes can be derived from the frequency positions of lines, and from their relative intensities, respectively. From the local force constants local static relaxations can be estimated. Because these relaxations affect the overlap between the electronic wave function of the defect and its nearest neighbors, they are important in a calculation of the electronic energy of a defect.

From the defect induced band mode absorption, and from localized mode sidebands, information on host lattice phonons, their densities, interactions, and perturbations is obtained.

7. Appendix

Table A1. Highest band and gap edge frequencies for alkali halides[a]

Substance	ω_{max} [cm^{-1}]	Gap edge freq.		T [K]	References
		lower	upper		
LiF	657	–	–		[112]
NaF	427	–	–	295	[124]
NaCl	261	–	–	80	[113]
NaBr	207	105	127	295	[114]
NaI	170	77	117	100	[115]
KCl	212	–	–	80	[116]
KBr	167	94	102	90	[115]
KI	142	69	98	95	[117]
RbF	283	125	163	80	[118]
RbCl	173	100	113	80	[118]
RbBr	130	–	–	80	[119]
RbI	105	–	–	80	[120]
CsCl	163	86	92	78	[121]
CsBr	112	–	–	80	[122]
CsI	84	–	–	295	[123]

[a] Values derived from Inelastic Neutron Scattering Experiments.

Acknowledgements. I am indebted to P. Grosse for his encouragement in writing this review, and would like to further thank him, together with H. Bilz, L. Genzel, R. Hayes, T. P. Martin, H. Pick, and R. Zeyher, for several stimulating discussions. I am also grateful to D. B. Fitchen for preprints of his recent papers and for allowing me to incorporate them in this review.

References

1. Fowler, W. B.: Color Centers in Alkali Halides. New York: Academic Press 1968.
2. Pick, H.: Optical Properties of Solids. Ed. Abelès, F. Amsterdam: North-Holland Publ. Comp. 1972.
3. Pick, H.: Struktur von Störstellen in Alkalihalogenidkristallen. Springer Tracts, Vol. 38. Berlin-Heidelberg-New York: Springer 1965.
4. Schulman, J. H., Compton, W. D.: Color Centers in Solids. New York: Pergamon Press 1963.
5. Elliott, R. J., Hayes, W., Jones, G. D., Macdonald, H. F., Sennett, C. T.: Proc. Roy. Soc. **289**, 1 (1965).
6. Bessent, R. G., Hayes, W., Hodby, I. W.: Phys. Letters **15**, 115 (1965).
7. Jones, G. D., Peled, S., Rosenwaks, S., Yatsiv, S.: Phys. Rev. **183**, 353 (1969).
8. Loudon, R.: Proc. Phys. Soc. (London) **84**, 379 (1964).
9. Loudon, R.: Advan. Phys. **13**, 432 (1964).
10. Knox, R. S., Gold, A.: Symmetry in the Solid State. New York Benjamin 1964.
11. Tinkham, M.: Group Theory and Quantum Mechanics. New York: MacGraw Hill 1964.
12. Mulliken, R. S.: J. Chem. Phys. **3**, 375 (1935).
13. Bouckaert, L. P., Smoluchowski, R., Wigner, E.: Phys. Rev. **50**, 58 (1936).
14. Bethe, H. A.: Ann. Phys. **3**, 133 (1929).
15. von der Lage, F. C., Bethe, H. A.: Phys. Rev. **71**, 619 (1947).
16. Howarth, D. J., Jones, H.: Proc. Phys. Soc. (London) A **65** 353 (1952).
17. Genzel, L., Festkörperprobleme, Vol., VI. London: Pergamon 1966.
18. Sievers, A. J.: Elementary Excitations in Solids, p. 193. New York: Plenum Press 1969.
19. Benedek, G., Nardelli, G. F.: Phys. Rev. **155**, 1004 (1967).
20. Klein, M. V.: see Ref. [1].
21. Hübner, R.: Z. Physik **222**, 380 (1969).
22. Kühner, D., Wagner, M.: Z. Physik **207**, 111, (1968).
23. Gethins, T., Timusk, T., Woll, E. J.: Phys. Rev. **157**, 744 (1967).
24. Sennet, C. T.: J. Phys. Chem. Solids **26**, 1097 (1965).
25. Xinh, N. X., Maradudin, A. A., Coldwell-Horsfall, R. A.: J. Phys. Radium **26**, 717 (1965).
26. Maradudin, A. A.: Solid State Phys., Vol. 19. Ed. Seitz, F., and Turnbull, D. New York: Academic Press 1966.
27. Benedek, G., Nardelli, G. F.: Phys. Rev. **154**, 872 (1967).
28. Gebhardt, W., Maier, K.: Phys. Stat. Sol. **8**, 303 (1965).
29. Rosenstock, H., Klick, C. C.: Phys. Rev. **119**, 1198 (1960).
30. Schäfer, G.: J. Phys. Chem. Sol. **12**, 233 (1960).
31. Price, W. C., Wilkinson, G. R.: Final Technical Report Nr. 2 (dec, 1960), US Army Contract No DA–91–591–EUC 1908, OI–4201–60 (R. + D. 260).
32. Mitsuishi, A., Yoshinaga, H.: Progr. Theoret. Phys. (Kyoto) Suppl. **23**, 241 (1962).
33. Mirlin, D. N., Reshina, I., I.: Sov. Phys. Sol. State **6**, 728 (1964).
34. Mirlin, D. N., Reshina, I. I.: Sov. Phys. Sol. State **6**, 2454 (1965).
35. Mirlin, D. N., Reshina, N. N.: Sov. Phys. Solid State **8**, 116 (1966).
36. Mitra, S. S., Brada, Y.: Phys. Letters **17**, 19 (1965).
37. Fritz, B., Groß, U., Bäuerle, D.: Phys. Stat. Sol. **11**, 231 (1965).
38. Dötsch, H., Gebhardt, W., Martius, C. H.: Phys. Stat. Sol. **3**, 9 (1965).

39. Mitra, S. S., Singh, R. R.: Phys. Rev. Letters 16, 694 (1966).
40. Timusk, T., Klein, M. V.: Phys. Rev. 141, 664 (1966).
41. Bäuerle, D., Fritz, B.: Phys. Stat. Sol. 24, 207 (1967).
42. Barth, W., Fritz, B.: Phys. Stat. Sol. 19, 515 (1967).
43. Fritz, B., Gerlach, J., Groß, U.: Localized Excitations in Solids. Ed. Wallis, R. F. New York: Plenum Press 1968.
44. Dötsch, H.: Phys. Stat. Sol. 31, 649 (1969).
45. Dötsch, H., Mitra, S. S.: Phys. Rev. 178, 1492 (1969).
46. MacPherson, R. W., Timusk, T.: Can. J. Phys. 48, 2146 (1970).
47. de Souza, M., Gongora, A. D., Aegerter, M., Lüty, F.: Phys. Rev. Letters 25, 1426 (1970).
48. Kurz, G., Susman, S., Internat. Conf. "Color Centers in Ionic Crystals", Reading, U. K., 1971.
49. Olson, C. G., Lynch, D. W.: Phys. Rev. 4, 1990, (1971).
50. Wallis, R., Maraduin, A. A.: Progr. Theor. Phys. (Kyoto) 24, 1055 (1960).
51. Bilz, H., Strauch, D., Fritz, B.: J. Phys. (Paris) 21. Suppl. C 2 − 3 (1966).
52. Bilz, H., Zeyher, R., Wehner, R. K.: Phys. Stat. Sol. 20, K 167 (1967).
53. Page, J. B., Dick, B. G.: Phys. Rev. 163, 910 (1967).
54. Xinh, N. X.: Phys. Rev. 163, 896 (1967).
55. Page, J. B., Strauch, D.: Phys. Stat. Sol. 24, 469 (1967).
56. Strauch, D., Page, J. B.: Phys. Stat. Sol. 30, 495 (1968).
57. Cunningham, S. L., Hardy, J. R.: Solid State Commun. 6, 769 (1968).
58. Strauch, D.: Phys. Stat. Sol. 30, 495 (1968).
59. Zeyher, R., Bilz, H., see Ref. [43], p. 767.
60. Zeyher, R., Bilz, H.: Phys. Stat. Sol. 31, 157 (1969).
61. Striefler, M., Jaswal, S. S.: J. Phys. Chem. Solids 30, 827 (1969).
62. Haridasan, T. M., Krishnamurthy, N.: J. Indian Inst. Sci. 51, 1 (1969).
63. Boese, F. K., Wagner, M.: Z. Physik 235, 140 (1970).
64. Lannoo, M., Dobrzynski, L.: J. Phys. Chem. Solids 33, 1447 (1972).
65. Harrington, J. A., Walker, C. T.: Phys. Stat. Sol. (b) 43, 619 (1971).
66. Harrington, J. A., Harley, R. T., Walker, C. T.: Solid State Commun. 9, 683 (1971).
67. Harrington, J. A., Weber, R.: Solid State Commun. 11, 1435 (1972).
68. Fritz, B., see Ref. [43].
69. Fritz, B.: J. Phys. Chem. Sol. 23, 375 (1962).
70. Gross, U., Bron, W. E.: Phys. Letters 25A, 312 (1967).
71. Bäuerle, D., Fritz, B.: Phys. Stat. Sol. 29, 639 (1968).
72. Dürr, U., Bäuerle, D.: Z. Physik 233, 94 (1970).
73. Akhvlediani, Z. G., Politov, N. G.: ZHETF, 10, 249 (1969).
74. Politov, N. G., Akhvlediani, Z. G., see Ref. [48].
75. Shamu, R. E., Hartmann, W. M., Yasaitis, E. L.: Phys. Rev. 170, 822 (1968).
76. Montgomery, G. P., Fenner, W. R., Klein, M. V., Timusk, T.: Phys. Rev. B 5, 3343 (1972).
77. Harrington, J. A., Harley, R. T., Walker, C. T.: Solid State Comm. 8, 407 (1970).
78. Hayes, W., Macdonald, H. F.: Proc. Roy. Soc. A 297, 503 (1967).
79. Jones, G. D., Satten, R. A.: Phys. Rev. 147, 566 (1966).
80. Dötsch, H.: Ph. D. Thesis, University of Frankfurt, 1967.
81. Lee, L. C., Faust, W. L.: Phys. Rev. Letters 26, 648 (1971).
82. Hayes, W., Macdonald, H. F., Elliott, R.: J. Phys. Rev. Letters 15, 961 (1965).
83. Cowley, R. A.: Advan. Phys. 12, 421 (1963).
84. Hurrell, J. P., Minkiewiez, V. J.: Solid State Commun. 8, 463 (1970).
85. Benedek, G., Nardelli, G. F.: Phys. Rev. Letters 16, 517 (1966).
86. Bäuerle, D.: Ph. D. Thesis, University of Stuttgart, 1969.
87. Bäuerle, D., Hübner, R.: Phys. Rev. B 2, 4252 (1970).
88. de Souza, M. F. Lüty, F.: see Ref. [48].
89. Woll, E. J., Gethins, T., Timusk, T.: Can. J. Phys. 46, 2263 (1968).

90. Bäuerle, D.: Internat. Symposium "Color Centers in Alkali Halides", Rom, 1968.
91. Bäuerle, D., Fritz, B.: Solid State Commun. **6**, 453 (1968).
92. Ram, P. N., Agrawal, B. K.: Solid State Commun. **11**, 1719 (1972).
93. Markham, J. J.: Solid State Phys. Suppl. **8**, ed. Seitz, F. and Turnbull, D. New York: Academic Press 1965.
94. Benedek, G., Mulazzi, E.: Phys. Rev. **179**, 906 (1969).
95. Singh, R., Mitra, S. S.: Phys. Rev. B **2**, 1070 (1970).
96. See e.g. Dexter, D. L.: Solid State Physics. Vol. 6, p. 353. ed. Seitz, F., Turnbull, D. New York: Academic Press 1958.
97. Jaswal, S. S.: Phys. Rev. **140**, A 687 (1965).
98. Ward, R. W., Timusk, T.: Phys. Rev. B **5**, 2331 (1972).
99. Kersten, R.: Ph. D. Thesis, University of Stuttgart, 1970.
100. Gnaedinger, R. J.: J. Chem. Phys. **21**, 323 (1953).
101. Martin, R. M.: Light Scattering in Solids. Ed. Balkanski, M. Paris: Flamarion Sci. 1971.
102. Fitchen, D. B., Buchenauer, C. J.: Physics of Impurity Centers in Crystals, Tallin, 1972, p. 277.
103. Wolfram, G., Jaswal, S. S., Sharma, T. P.: Phys. Rev. Letters **29**, 160 (1972).
104. Worlock, J. M., Porto, S. P. S.: Phys. Rev. Letters **15**, 697 (1965).
105. Buchenauer, C. J., Fitchen, D. B., Page, J. B.: Light Scattering Spectra of Solids, p. 521. Ed. Wright, G. B. Berlin-Heidelberg-New York: Springer 1969.
106. Bendek, G., Mulazzi, E.: see Ref. [105], p. 531.
107. Fritz, B.: see Ref. [43], p. 496.
108. Henry, C. H., Slichter, C. P.: see Ref. [3], p. 351.
109. Rebane, K., Hizhnyakov, V., Tehver, I.: Izv. Akad. Nauk. Est. SSR, Ser. Fiz.-Mat. Tekh. Nauk. **16**, 207 (1967).
110. Fitchen, D. B., Buchenauer, C. J.: see Ref. [48].
111. Mieher, R. L.: Phys. Rev. Letters **8**, 362 (1962).
112. Dolling, G., Smith, H. G., Nicklow, R. M., Vijayaraghavan, P. R., Wilkinson, M. K.: Phys. Rev. **168**, 970 (1968).
113. Raunio, G., Rolandson, S.: Phys. Rev. B **2**, 2098 (1970).
114. Reid, J. S., Smith, T., Buyers, W. J. L.: Phys. Rev. B **1**, 1833 (1970).
115. Cowley, R. A., Cochran, W., Brockhouse, B. N., Woods, A. D. B.: Phys. Rev. **119**, 980 (1960).
116. Raunio, G., Almqvist, L.: Phys. Stat. Sol. **33**, 209 (1969).
117. Dolling, G., Cowley, R. A., Schittenhelm, C., Thorson, I. M.: Phys. Rev. **147**, 577 (1966).
118. Raunio, G., Rolandson, S.: J. Phys. C **3**, 1013 (1970).
119. Rolandson, S., Raunio, G.: J. Phys. C **4**, 958 (1971).
120. Raunio, G., Rolandson, S.: Phys. Stat. Sol. (b) **40**, 749 (1970).
121. Ahmad, A. A. Z., Smith, H. G., Wakabayashi, N., Wilkinson, M. K.: Phys. Rev. B **6**, 3956 (1972).
122. Rolandson, S., Raunio, G.: Phys. Rev. B **4**, 4617 (1971).
123. Bührer, W., Hälg, W.: Phys. Stat. Sol. (b) **46**, 679 (1971).
124. Buyers, W. J. L.: Phys. Rev. **153**, 923 (1967).

Dr. D. Bäuerle
Max-Planck-Institut für Festkörperforschung
7000 Stuttgart 1
Heilbronner Str. 69
and
Philips Forschungslaboratorium
5100 Aachen
Weißhausstr. 4
Germany